MECHANICAL ENGINEERING FORMULAS

Pocket Guide

Tyler G. Hicks, P.E.

International Engineering Associates
Member: American Society of Mechanical Engineers,
U.S. Naval Institute, and
Steamship Historical Society of America

McGRAW-HILL

New York   Chicago   San Francisco   Lisbon   London   Madrid
Mexico City   Milan   New Delhi   San Juan   Seoul
Singapore   Sydney   Toronto

The *Mcgraw·Hill* Companies

Library of Congress Cataloging-in-Publication Data

Hicks, Tyler Gregory, date.
  Mechanical engineering formulas pocket guide / Tyler G. Hicks.

      p.   cm.
  Includes index.
  ISBN 0-07-135609-6
    1. Mechanical engineering—Handbooks, manuals, etc.   I. Title.

  TJ151 .H62   2003
  621—dc21

                                        2002040975

Copyright © 2003 by The McGraw-Hill Companies, Inc. All rights reserved. Printed in the United States of America. Except as permitted under the United States Copyright Act of 1976, no part of this publication may be reproduced or distributed in any form or by any means, or stored in a data base or retrieval system, without the prior written permission of the publisher.

5 6 7 8 9 0   QSR/QSR   0 9

ISBN-13: 978-0-07135609-1
ISBN-10: 0-07-135609-6

The sponsoring editor for this book was Larry S. Hager, the editing supervisor was David E. Fogarty, and the production supervisor was Pamela A. Pelton. It was set in Times Roman by Pro-Image Corporation.

Printed and bound by Quebecor World.

This book was printed on recycled, acid-free paper containing a minimum of 50% recycled, de-inked fiber.

McGraw-Hill books are available at special quantity discounts to use as premiums and sales promotions, or for use in corporate training programs. For more information, please write to the Director of Special Sales, Professional Publishing, McGraw-Hill, Two Penn Plaza, New York, NY 10121-2298. Or contact your local bookstore.

Information contained in this work has been obtained by The McGraw-Hill Companies, Inc. ("McGraw-Hill") from sources believed to be reliable. However, neither McGraw-Hill nor its authors guarantee the accuracy or completeness of any information published herein, and neither McGraw-Hill nor its authors shall be responsible for any errors, omissions, or damages arising out of use of this information. This work is published with the understanding that McGraw-Hill and its authors are supplying information but are not attempting to render engineering or other professional services. If such services are required, the assistance of an appropriate professional should be sought.

CONTENTS

Section 5. Machine Component and Reliability Formulas   77

Section 6. Metalworking Formulas   137

Section 7. Formulas for Heating, Ventilating, and Air Conditioning

Section 8. Thermodynamics Formulas

Section 9. Energy Engineering Formulas

Section 10. Formulas for Fluids Engineering   **227**

PREFACE

This handy book presents some 2000 needed formulas for mechanical engineers to help them in the design office, in the field, and on a variety of consulting jobs, anywhere in the world. These formulas are also useful to design drafters, product engineers, job estimators, stress analysis engineers, machine designers, professional-engineer license candidates, civil service examination candidates, and many other people in a variety of mechanical engineering pursuits.

The book presents formulas in 10 different specialties of mechanical engineering: mechanics; strength of materials; machine components; shafts and shafting; metalworking; heating, ventilating, and air conditioning; thermodynamics; energy engineering; fluids engineering; and conversion factors for U.S. Customary System (USCS) units and International System (*Système International,* or SI) units. Key formulas are presented for each of these topics. Each formula is explained so the engineer, drafter, or designer knows how, where, and when to use it in professional work. Most of the formulas are given in both USCS and SI units. Hence, the book is usable throughout the world. To assist the mechanical engineer using this material in worldwide engineering practice, a comprehensive tabulation of conversion factors is presented in Section 1.

In assembling this collection of formulas, the author was guided by experts who recommended the areas of greatest need for a handy book of practical and applied mechanical engineering formulas.

Sources for the formulas presented here include the various regulatory and industry groups in the field of mechanical engineering, authors of recognized authoritative books on important topics in the field, drafters, researchers in the field of mechanical engineering, and a number of design engineers in the field of product engineering. These sources are cited in the Acknowledgments section of this book.

When employing any of the formulas in this book that may come from an industry or regulatory code, the user is cautioned to consult the latest version of the code. Formulas may be changed from one edition to the next. In a work of this magnitude it is difficult to include the latest formulas from the numerous constantly changing codes. Hence, the formulas given here are those current at the time of publication of this book.

In a work this large errors may occur. Hence, the author will be grateful to any reader who detects an error and calls it to the author's attention. Just write the author in care of the publisher. The error will be corrected in the next printing.

In addition, if a user believes that one or more important formulas have been left out, the author will be happy to consider them for inclusion in the next edition of the book. Again, just write him in care of the publisher.

Tyler G. Hicks, P.E.

ACKNOWLEDGMENTS

In preparing the mechanical engineering formulas for this handy book, the author had the help and guidance of a number of engineering experts in the field. In addition, he consulted many authoritative published sources. Hence, the author wishes to acknowledge this assistance and give thanks to those individuals, organizations and publications which were of help to him in compiling this collection of useful formulas.

Published sources consulted by the author include engineering society standards—namely those available from the American Society of Mechanical Engineers, American Society of Tool and Manufacturing Engineers, and the American Society of Civil Engineers. Other authoritative published sources include a number of published engineering books, namely: Wahl, *Mechanical Springs;* Carrier Air Conditioning Corporation, *Handbook of Air Conditioning System Design;* Nordhoff, *Machine Shop Estimating;* Hicks, *Standard Handbook of Engineering Calculations;* Yeaple, *Hydraulic and Pneumatic Power and Control;* Greenwood, *Engineering Data for Product Design;* Bell, Jr., *HVAC Equations, Data, and Rules of Thumb;* Grimm and Rosaler, *Handbook of HVAC Design;* U.S. Department of the Interior, Bureau of Reclamation, *Metric Manual;* plus hundreds of others.

The author also thanks Larry Hager, Senior Editor, Professional Group, The McGraw-Hill Companies, for his excellent guidance and patience during the long preparation of the manuscript for this book. The author gives gracious thanks to his wife, Mary Shanley Hicks, a publishing professional, who always most willingly offered help and advice when needed.

The author is grateful to the writers of all the publications cited and consulted, for the insight that they gave him into mechanical engineering formulas.

Tyler G. Hicks, P.E.

HOW TO USE THIS BOOK

The formulas presented in this book are intended for use by mechanical engineers in every aspect of their professional work—machinery design, evaluation, testing, construction, specification, repair, etc.

To find a suitable formula for the situation you face, start by consulting the index. Every effort has been made to present a comprehensive listing of all formulas in the book.

Once you find the formula you seek, read any accompanying text giving background information about the formula. Solve the formula and use the results for the task at hand.

Where a formula may come from a regulatory code, or where a code exists for the particular work being done, be certain to check the latest edition of the applicable code to see that the given formula agrees with the code formula. If it does not agree, be certain to use the latest code formula available. Remember, as a design engineer, you are responsible for the machinery you plan, design, and build. Using the latest edition of any governing code is the only sensible way to produce a safe and dependable design that you will be proud to be associated with. Further, you will sleep more peacefully!

SECTION 1
CONVERSION FACTORS FOR MECHANICAL ENGINEERING PRACTICE

Mechanical engineers throughout the world accept both the *United States Customary System* (USCS) and the International System (*Système International,* or SI) units of measure for both applied and theoretical calculations. However, the SI units are much more widely used than those of the USCS. Hence, both the USCS and the SI units are presented for essentially every formula in this book. Thus, the user of the book can apply the formulas with confidence anywhere in the world.

To permit even wider use of this text, this section contains the conversion factors needed to switch from one system to the other. For engineers unfamiliar with either system of units, the author suggests the following steps for becoming acquainted with the unknown system:

1. Prepare a list of measurements commonly used in your daily work.

2. Insert, opposite each known unit, the unit from the other system. Table 1.1 shows such a list of USCS units with corresponding SI units and symbols prepared by a mechanical engineer who normally uses the USCS. The SI units shown in Table 1.1 were obtained from Table 1.3 by the engineer.

3. Find, from a table of conversion factors, such as Table 1.3, the value used to convert from USCS to SI units. Insert each appropriate value in Table 1.2 from Table 1.3.

4. Apply the conversion values wherever necessary for the formulas in this book.

5. Recognize—here and now—that the most difficult aspect of becoming familiar with a new system of measurement is to become comfortable with the names and magnitudes of the units. Numerical conversion is simple, once you have set up your own conversion table.

TABLE 1.1 Commonly Used USCS and SI Units†

USCS unit	SI unit	SI symbol	Conversion factor—multiply USCS unit by this factor to obtain the SI unit
square foot	square meter	m²	0.929
cubic foot	cubic meter	m³	0.2831
pounds per square inch	kilopascal	kPa	6.894
pound force	newton	N	4.448
foot-pound torque	newton-meter	N · m	1.356
Btu per pound	kilojoules per kilogram	kJ/kg	2.326
gallons per minute	liters per second	L/s	0.06309
Btu per cubic foot	kilojoules per cubic meter	kJ/m³	37.26

†Because of space limitations this table is abbreviated. For a typical engineering practice an actual table would be many times this length.

TABLE 1.2 Typical Conversion Table†

To convert from	To	Multiply by
square feet	square meters	9.290304 E − 02
feet per second squared	meters per second squared	3.048 E − 01
cubic feet	cubic meters	2.831685 E − 02
pounds per cubic inch	kilogram per cubic meter	2.767990 E + 04
gallons per minute	liters per second	6.309 E − 02
pounds per square inch	kilopascals	6.894757
pound force	newtons	4.448222
British thermal units per square foot	joules per square meter	1.135653 E + 04
British thermal units per hour	watts	2.930711 E − 01
British thermal units per cubic foot	kilojoules per cubic meter	3.725697 E + 01
foot-pounds (torque)	newton-meters	1.355818
British thermal units per pound	joules per kilogram	2.326 E + 03

Note: The E indicates an exponent, as in scientific notation, followed by a positive or negative number, representing the power of 10 by which the given conversion factor is to be multiplied before use. Thus, for the square feet conversion factor, 9.290304 × ¹⁄₁₀₀ = 0.09290304, the factor to be used to convert square feet to square meters. For a positive exponent, as in converting British thermal units per cubic foot to kilojoules per cubic meter, 3.725697 × 10 = 37.25697.

Where a conversion factor cannot be found, simply use the dimensional substitution. Thus, to convert pounds per cubic inch to kilograms per cubic meter, find 1 lb = 0.4535924 kg, and 1 in³ = 0.00001638706 m³. Then, 1 lb/in³ = 0.4535924 kg/0.00001638706 m³ = 27,67990, or 2.767990 E + 04.

†This table contains only selected values. See the U.S. Department of the Interior, *Metric Manual,* or National Bureau of Standards, *The International System of Units (SI),* both available from the Government Printing Office (GPO), for far more comprehensive listings of conversion factors.

Be careful, when you are using formulas containing a numerical constant, to convert the constant to that value for the system you are using. You can, however, use the formula for the USCS units (when the formula is given in those units) and then convert the final result to the SI equivalent by using Table 1.3. For the few formulas given in SI units, the reverse procedure should be used.

TABLE 1.3 Factors for Conversion to SI Units of Measurement

To convert from	To	Multiply by
acre-foot acre · ft	cubic meter, m^3	1.233489 E + 03
acre	square meter, m^2	4.046873 E + 03
angstrom, Å	meter, m	1.000000† E − 10
atmosphere, atm (standard)	pascal, Pa	1.013250† E + 05
atmosphere, atm (technical = 1 kgf/cm^2)	pascal, Pa	9.806650† E + 04
bar	pascal, Pa	1.000000† E + 05
barrel (for petroleum, 42 gal)	cubic meter, m^3	1.589873 E − 01
board foot, board ft	cubic meter, m^3	2.359737 E − 03
British thermal unit, Btu (mean)	joule, J	1.05587 E + 03
British thermal unit, Btu (International Table) · in/(h)(ft^2) (°F) (k, thermal conductivity)	watt per meter-kelvin, $W/(m \cdot K)$	1.442279 E − 01
British thermal unit, Btu (International Table)/h	watt, W	2.930711 E − 01
British thermal unit, Btu (International Table)/ (h)(ft^2)(°F) (C, thermal conductance)	watt per square meter-kelvin, $W/(m^2 \cdot K)$	5.678263 E + 00
British thermal unit, Btu (International Table)/lb	joule per kilogram, J/kg	2.326000† E + 03
British thermal unit, Btu (International Table)/(lb)(°F) (c, heat capacity)	joule per kilogram-kelvin, $J/(kg \cdot K)$	4.186800† E + 03
British thermal unit per cubic foot, Btu (International Table)/ft^3	joule per cubic meter, J/m^3	3.725895 E + 04
bushel (U.S.)	cubic meter, m^3	3.523907 E − 02

TABLE 1.3 Factors for Converson to SI Units of Measurement (*Continued*)

To convert from	To	Multiply by
calorie (mean)	joule, J	4.19002 E + 00
candela per square inch, cd/in²	candela per square meter, cd/m²	1.550003 E + 03
centimeter, cm, of mercury (0°C)	pascal, Pa	1.33322 E + 03
centimeter, cm, of water (4°C)	pascal, Pa	9.80638 E + 01
chain	meter, m	2.011684 E + 01
circular mil	square meter, m²	5.067075 E − 10
day	second, s	8.640000† E + 04
day (sidereal)	second, s	8.616409 E + 04
degree (angle)	radian, rad	1.745329 E − 02
degree Celsius	kelvin, K	$T_K = t_C + 273.15$
degree Fahrenheit	degree Celsius, °C	$t_C = (t_F - 32)/1.8$
degree Fahrenheit	kelvin, K	$T_K = (t_F + 459.67)/1.8$
degree Rankine	kelvin, K	$T_K = T_R/1.8$
(°F)(h)(ft²)/Btu (International Table) (*R*, thermal resistance)	kelvin-square meter per watt, K · m²/W	1.761102 E − 01
(°F)(h)(ft²)/[Btu (International Table) · in] (thermal resistivity)	kelvin-meter per watt, K · m/W	6.933471 E + 00
dyne, dyn	newton, N	1.000000† E − 05
fathom	meter, m	1.828804 E + 00
foot, ft	meter, m	3.048000† E − 01
foot, ft (U.S. survey)	meter, m	3.048006 E − 01
foot, ft, of water (39.2°F) (pressure)	pascal, Pa	2.98898 E + 03
square foot, ft²	square meter, m²	9.290304† E − 02
square foot per hour, ft²/h (thermal diffusivity)	square meter per second, m²/s	2.580640† E − 05
square foot per second, ft²/s	square meter per second, m²/s	9.290304† E − 02
cubic foot, ft³ (volume or section modulus)	cubic meter, m³	2.831685 E − 02
cubic foot per minute, ft³/min	cubic meter per second, m³/s	4.719474 E − 04
cubic foot per second, ft³/s	cubic meter per second, m³/s	2.831685 E − 02

TABLE 1.3 Factors for Converson to SI Units of Measurement (*Continued*)

To convert from	To	Multiply by
foot to the fourth power, ft^4 (area moment of inertia)	meter to fourth power, m^4	8.630975 E − 03
foot per minute, ft/min	meter per second, m/s	5.080000† E − 03
foot per second, ft/s	meter per second, m/s	3.048000† E − 01
foot per second squared, ft/s^2	meter per second squared, m/s^2	3.048000† E − 01
footcandle, fc	lux, lx	1.076391 E + 01
foot-lambert, ft · L	candela per square meter, cd/m^2	3.426259 E + 00
foot-pound force, ft · lbf	joule, J	1.355818 E + 00
foot-pound force per minute, ft · lbf/min	watt, W	2.259697 E − 02
foot-pound force per second, ft · lbf/s	watt, W	1.355818 E + 00
foot poundal, ft poundal	joule, J	4.214011 E − 02
free fall, standard g	meter per second squared, m/s^2	9.806650† E + 00
gallon, gal (Canadian liquid)	cubic meter, m^3	4.546090 E − 03
gallon, gal (U.K. liquid)	cubic meter, m^3	4.546092 E − 03
gallon, gal (U.S. dry)	cubic meter, m^3	4.404884 E − 03
gallon, gal (U.S. liquid)	cubic meter, m^3	3.785412 E − 03
gallon, gal (U.S. liquid) per day	cubic meter per second, m^3/s	4.381264 E − 08
gallon, gal (U.S. liquid) per minute	cubic meter per second, m^3/s	6.309020 E − 05
grad	degree (angular)	9.000000† E − 01
grad	radian, rad	1.570796 E − 02
grain, gr	kilogram, kg	6.479891† E − 05
gram, g	kilogram, kg	1.000000† E − 03
hectare, ha	square meter, m^2	1.000000† E + 04
horsepower, hp (550 ft · lbf/s)	watt, W	7.456999 E + 02
horsepower, hp (boiler)	watt, W	9.80950 E + 03
horsepower, hp (electric)	watt, W	7.460000† E + 02
horsepower, hp (water)	watt, W	7.46043† E + 02
horsepower, hp (U.K.)	watt, W	7.4570 E + 02

TABLE 1.3 Factors for Converson to SI Units of Measurement (*Continued*)

To convert from	To	Multiply by
hour, h	second, s	3.600000† E + 03
hour, h (sidereal)	second, s	3.590170 E + 03
inch, in	meter, m	2.540000† E − 02
inch of mercury, inHg (32°F) (pressure)	pascal, Pa	3.38638 E + 03
inch of mercury, inHg (60°F) (pressure)	pascal, Pa	3.37685 E + 03
inch of water inH$_2$O (60°F) (pressure)	pascal, Pa	2.4884 E + 02
square inch, in^2	square meter, m^2	6.451600† E − 04
cubic inch, in^3 (volume or section modulus)	cubic meter, m^3	1.638706 E − 05
inch to the fourth power, in^4 (area moment of inertia)	meter to fourth power, m^4	4.162314 E − 07
inch per second, in/s	meter per second, m/s	2.540000† E − 02
kelvin, K	degree Celsius, °C	$t_C = T_K − 273.15$
kilogram force, kgf	newton, N	9.806650† E + 00
kilogram force-meter, kg · m	newton-meter, N · m	9.806650† E + 00
kilogram force-second squared per meter, kgf · s^2/m (mass)	kilogram, kg	9.806650† E + 00
kilogram force per square centimeter, kgf/cm^2	pascal, Pa	9.806650† E + 04
kilogram force per square meter, kgf/m^2	pascal, Pa	9.806650† E + 00
kilogram force per square millimeter, kgf/mm^2	pascal, Pa	9.806650† E + 06
kilometer per hour, km/h	meter per second, m/s	2.777778 E − 01
kilowatthour, kWh	joule, J	3.600000† E + 06
kip (1000 lbf)	newton, N	4.448222 E + 03
kip per square inch, kip/in^2 or ksi	pascal, Pa	6.894757 E + 06
knot, kn (international)	meter per second, m/s	5.144444 E − 01
lambert, L	candela per square meter, cd/m^2	3.183099 E + 03

TABLE 1.3 Factors for Converson to SI Units of Measurement (*Continued*)

To convert from	To	Multiply by
liter	cubic meter, m³	1.000000† E − 03
maxwell	weber, Wb	1.000000† E − 08
mho	siemens, S	1.000000† E + 00
microinch, μin	meter, m	2.540000† E − 08
micrometer, μm	meter, m	1.000000† E − 06
miles, mi	meter, m	2.540000† E − 05
mile, mi (international)	meter, m	1.609344† E + 03
mile, mi (U.S statute)	meter, m	1.609347 E + 03
mile, mi (international nautical)	meter, m	1.852000† E + 03
mile, mi (U.S. nautical)	meter, m	1.852000† E + 03
square mile, mi² (international)	square meter, m²	2.589988 E + 06
square mile, mi² (U.S. statute)	square meter, m²	2.589998 E + 06
mile per hour, mi/h (international)	meter per second, m/s	4.470400† E − 01
mile per hour, mi/h (international)	kilometer per hour, km/h	1.609344† E + 00
millibar, mbar	pascal, Pa	1.000000† E + 02
millimeter of mercury, mmHg (0°C)	pascal, Pa	1.33322 E + 02
minute (angle)	radian, rad	2.908882 E − 04
minute, min	second s	6.000000† E + 01
minute (sidereal)	second, s	5.983617 E + 01
ounce, oz (avoirdupois)	kilogram, kg	2.834952 E − 02
ounce, oz (troy or apothecary)	kilogram, kg	3.110348 E − 02
ounce, oz (U.K. fluid)	cubic meter, m³	2.841307 E − 05
ounce, oz (U.S. fluid)	cubic meter, m³	2.957353 E − 05
ounce force, ozf	newton, N	2.780139 E − 01
ounce force-inch, ozf · in	newton-meter, N · m	7.061552 E − 03
ounce per square foot, oz (avoirdupois)/ft²	kilogram per square meter, kg/m²	3.051517 E − 01
ounce per square yard, oz (avoirdupois)/yd²	kilogram per square meter, kg/m²	3.390575 E − 02

TABLE 1.3 Factors for Converson to SI Units of Measurement (*Continued*)

To convert from	To	Multiply by	
perm (0°C)	kilogram per pascal-second-meter, kg/(Pa · s · m)	5.72135	E − 11
perm (23°C)	kilogram per pascal-second-meter, kg/(Pa · s · m)	5.74525	E − 11
perm-inch, perm · in (0°C)	kilogram per pascal-second-meter, kg/(Pa · s · m)	1.45322	E − 12
perm-inch, perm · in (23°C)	kilogram per pascal-second-meter, kg/(Pa · s · m)	1.45929	E − 12
pint, pt (U.S. dry)	cubic meter, m^3	5.506105	E − 04
pint, pt (U.S. liquid)	cubic meter, m^3	4.731765	E − 04
poise, P (absolute viscosity)	pascal-second, Pa · s	1.000000†	E − 01
pound, lb (avoirdupois)	kilogram, kg	4.535924	E − 01
pound, lb (troy or apothecary)	kilogram, kg	3.732417	E − 01
pound-square inch, lb · in^2 (moment of inertia)	kilogram-square meter, kg · m^2	2.926397	E − 04
pound per foot-second, lb/ft · s	pascal-second, Pa · s	1.488164	E + 00
pound per square foot, lb/ft^2	kilogram per square meter, kg/m^2	4.882428	E + 00
pound per cubic foot, lb/ft^3	kilogram per cubic meter, kg/m^3	1.601846	E − 01
pound per gallon, lb/gal (U.K. liquid)	kilogram per cubic meter, kg/m^3	9.977633	E + 01
pound per gallon lb/gal (U.S. liquid)	kilogram per cubic meter, kg/m^3	1.198264	E + 02
pound per hour, lb/h	kilogram per second, kg/s	1.259979	E − 04
pound per cubic inch, lb/in^3	kilogram per cubic meter, kg/m^3	2.767990	E + 04
pound per minute, lb/min	kilogram per second, kg/s	7.559873	E − 03
pound per second, lb/s	kilogram per second, kg/s	4.535924	E − 01
pound per cubic yard, lb/yd^3	kilogram per cubic meter, kg/m^3	5.932764	E − 01

TABLE 1.3 Factors for Converson to SI Units of Measurement (*Continued*)

To convert from	To	Multiply by
poundal	newton, N	1.382550 E − 01
pound force, lbf	newton, N	4.448222 E + 00
pound force-foot, lbf · ft	newton-meter, N · m	1.355818 E + 00
pound force per foot, lbf/ft	newton per meter, N/m	1.459390 E + 01
pound force per square foot, lbf/ft²	pascal, Pa	4.788026 E + 01
pound force per inch, lbf/in	newton per meter, N/m	1.751268 E + 02
pound force per square inch, lbf/in² (psi)	pascal, Pa	6.894757 E + 03
quart, qt (U.S. dry)	cubic meter, m³	1.101221 E − 03
quart qt (U.S. liquid)	cubic meter, m³	9.463529 E − 04
rod	meter, m	5.029210 E + 00
second (angle)	radian, rad	4.848137 E − 06
second (sidereal)	second, s	9.972696 E − 01
square (100 ft²)	square meter, m²	9.290304† E + 00
ton (assay)	kilogram, kg	2.916667 E − 02
ton (long, 2240 lb)	kilogram, kg	1.016047 E + 03
ton (metric), t	kilogram, kg	1.000000† E + 03
ton (refrigeration)	watt, W	3.516800 E + 03
ton (register)	cubic meter, m³	2.831685 E + 00
ton (short 2000 lb)	kilogram, kg	9.071847 E + 02
ton (long) per cubic yard, ton/yd³	kilogram per cubic meter, kg/m³	1.328939 E + 03
ton (short) per cubic yard, ton/yd³	kilogram per cubic meter kg/m³	1.186553 E + 03
ton force (2000 lbf)	newton, N	8.896444 E + 03
tonne, t	kilogram, kg	1.000000† E + 03
watt hour, Wh	joule, J	3.600000† E + 03
yard, yd	meter, m	9.144000† E − 01
square yard, yd²	square meter, m²	8.361274 E − 01
cubic yard, yd³	cubic meter, m³	7.645549 E − 01
year (365 days), yr	second, s	3.153600† E + 07
year (sidereal)	second, s	3.155815 E + 07

† Exact value.
Source: E380, *Standard for Metric Practice,* American Society for Testing and Materials.

SECTION 2
MECHANICS—STATICS AND KINETICS FORMULAS

FORMULAS OF MOTION

Nomenclature

t = time, s

s = linear displacement, ft (m)

v = linear velocity, ft/s (m/s)

V_0 = linear velocity at time zero, ft/s (m/s)

a = linear acceleration, ft/s^2 (m/s^2)

θ = angular displacement, rad

ω = angular velocity, rad/s

ω_0 = angular velocity at time zero, rad/s

α = angular acceleration, rad/s^2

w = weight of body, lbm (kg mass)

f = force of acceleration, lb (N)

g_c = conversion factor = 32.2 (lbm)(ft)/(lbf $\cdot$ s^2) (9.81 m/s^2)

v = constant	ω = constant	v = variable	ω = variable
$v = s/t$	$\omega = \theta/t$	$v = ds/dt$	$\omega = d\theta/dt$

a = constant	α = constant	a = variable	α = variable
$v = V_0 + at$	$\omega = \omega_0 + \alpha t$	$a = \dfrac{dv}{dt} = \dfrac{d^2s}{dt^2}$	$\alpha = \dfrac{d\omega}{dt} = \dfrac{d^2\theta}{dt^2}$
$s = V_0 t + \frac{1}{2}at^2$	$\theta = \omega_0 t + \frac{1}{2}\alpha t^2$	$v = \int a\, dt$	$\omega = \int \alpha\, dt$
$v = \sqrt{V_0^2 + 2as}$	$\omega = \sqrt{\omega_0^2 + 2\alpha\theta}$	$s = \int v\, dt$	$\theta = \int \omega\, dt$

For uniform acceleration

$$f = \frac{w}{g_c} a$$

STATICS

Any force system in space will be in equilibrium if the resultant force and resultant moment are both equal to zero. This can be expressed by

$$\Sigma F_x = \Sigma F_y = \Sigma F_z = 0 \qquad (2.1a)$$

$$\Sigma M_x = \Sigma M_y = \Sigma M_z = 0 \qquad (2.1b)$$

where F = force, lb (N)
M = moment, ft·lb (N·m)
x, y, z = orthogonal axes

Moment of Inertia

Rectangular Moment of Inertia. The rectangular moment of inertia I (Table 2.1) with respect to a given axis is defined by the equations

$$I = \int y^2 \, dm \quad \text{lb·ft}^2 \ (\text{kg·m}^2) \qquad \text{for solid body} \qquad (2.2)$$

and $\quad I = \int y^2 \, dA \quad \text{ft}^4 \ (\text{m}^4) \qquad \text{for plane area} \qquad (2.3)$

where y = distance from elements of mass or area to reference axis, ft (m)
dm = element of mass, lb (kg)
dA = element of area, ft^2 (m^2)

Radius of Gyration. The radius of gyration is a length K, ft (m), such that

$$I = \int y^2 \, dm = K^2 m \qquad \text{for solid body} \qquad (2.4)$$

and $\quad I = \int y^2 \, dA = K^2 A \qquad \text{for plane area} \qquad (2.5)$

where m = total mass, lb (kg), and A = area, ft^2 (m^2).

Rectangular Moments of Inertia and Radii of Gyration about Parallel Axes. Moment of inertia about any reference axis is equal to

TABLE 2.1 Properties of Various Cross Sections

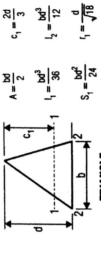

RECTANGLE

$$A = bd \qquad I_1 = \frac{bd^3}{12}$$

$$c_1 = d/2 \qquad I_2 = \frac{bd^3}{3}$$

$$c_2 = d \qquad I_3 = \frac{b^3 d^3}{6(b^2 + d^2)}$$

$$c_3 = \frac{bd}{\sqrt{b^2 + d^2}} \qquad S_3 = \frac{b^2 d^2}{6\sqrt{b^2 + d^2}}$$

$$S_1 = \frac{bd^2}{6} \qquad r_3 = \frac{bd}{\sqrt{6(b^2 + d^2)}}$$

$$r_1 = \frac{d}{\sqrt{12}}$$

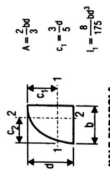

TRIANGLE

$$A = \frac{bd}{2} \qquad c_1 = \frac{2d}{3}$$

$$I_1 = \frac{bd^3}{36} \qquad I_2 = \frac{bd^3}{12}$$

$$S_1 = \frac{bd^2}{24} \qquad r_1 = \frac{d}{\sqrt{18}}$$

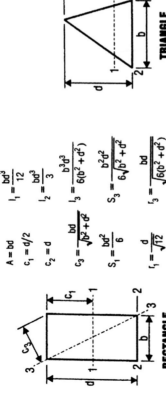

PARABOLA

$$A = \frac{2}{3}bd \qquad c = \frac{3}{5}d$$

$$I_1 = \frac{8}{175}bd^3 \qquad I_2 = \frac{b^3 d}{30}$$

$$I_3 = \frac{16}{105}bd$$

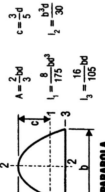

HALF PARABOLA

$$A = \frac{2}{3}bd \qquad c_2 = \frac{5}{8}b$$

$$c_1 = \frac{3}{5}d \qquad I_2 = \frac{19}{480}b^3 d$$

$$I_1 = \frac{8}{175}bd^3$$

13

TABLE 2.1 Properties of Various Cross Sections (*Continued*)

Section	Moment of inertia	Section modulus	Radius of gyration
Equilateral Polygon A = area R = rad circumscribed circle r = rad inscribed circle n = no. sides a = length of side Axis as in preceding section of octagon	$I = \dfrac{A}{24}(6R^2 - a^2)$ $= \dfrac{A}{48}(12r^2 + a^2)$ $= \dfrac{AR^2}{4}$ (approx)	$\dfrac{I}{c} = \dfrac{I}{r}$ $= \dfrac{I}{R\cos\dfrac{180°}{n}}$ $= \dfrac{AR}{4}$ (approx)	$\sqrt{\dfrac{6R^2 - a^2}{24}} \approx \dfrac{R}{2}$ $\sqrt{\dfrac{12r^2 + a^2}{48}}$
	$I = \dfrac{6b^2 + 6bb_1 + b_1{}^2}{36(2b + b_1)}h^3$ $c = \dfrac{1}{3}\dfrac{3b + 2b_1}{2b + b_1}h$	$\dfrac{I}{c} = \dfrac{6b^2 + 6bb_1 + b_1{}^2}{12(3b + 2b_1)}h^2$	$\dfrac{h\sqrt{12b^2 + 12bb_1 + 2b_1{}^2}}{6(2b + b_1)}$

14

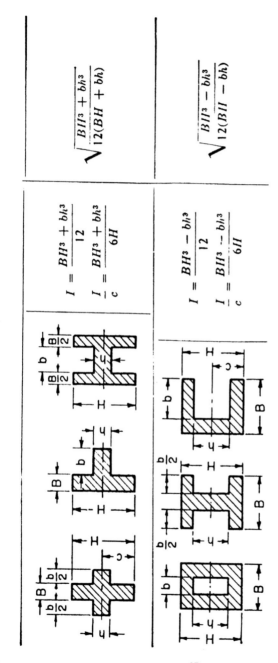

$$I = \frac{BH^3 + bh^3}{12}$$

$$\frac{I}{c} = \frac{BH^3 + bh^3}{6H}$$

$$\sqrt{\frac{BH^3 + bh^3}{12(BH + bh)}}$$

$$I = \frac{BH^3 - bh^3}{12}$$

$$\frac{I}{c} = \frac{BH^3 - bh^3}{6H}$$

$$\sqrt{\frac{BH^3 - bh^3}{12(BH - bh)}}$$

15

TABLE 2.1 Properties of Various Cross Sections (*Continued*)

Section	Moment of inertia and section modulus	Radius of gyration
	$$I = \tfrac{1}{3}(Bc_1{}^3 - B_1h^3 + bc_2{}^3 - b_1h_1{}^3)$$ $$c_1 = \frac{1}{2}\,\frac{aH^2 + B_1d^2 + b_1d_1(2H - d_1)}{aH + B_1d + b_1d_1}$$	$$\sqrt{\dfrac{I}{(Bd + bd_1) + a(h + h_1)}}$$
		$$I = \tfrac{1}{3}(Bc_1{}^3 - bh^3 + ac_2{}^3)$$ $$c_1 = \frac{1}{2}\,\frac{a\,li^2 + bd^2}{a\,ll + bd}$$ $$c_2 = ll - c_1$$ $$r = \sqrt{\dfrac{I}{[Bd + a(ll - d)]}}$$

Section	Moment of inertia	Section modulus	Radius of gyration
	$I = \dfrac{\pi d^4}{64} = \dfrac{\pi r^4}{4} = \dfrac{A}{4} r^2$ (approx) $= 0.05 d^4$ (approx)	$\dfrac{I}{c} = \dfrac{\pi d^3}{32} = \dfrac{\pi r^3}{4} = \dfrac{A}{4} r$ $= 0.1 d^3$ (approx)	$\dfrac{r}{2} = \dfrac{d}{4}$
$d_m = \tfrac{1}{2}(D + d)$ $s = \tfrac{1}{2}(D - d)$	$I = \dfrac{\pi}{64}(D^4 - d^4)$ $= \dfrac{\pi}{4}(R^4 - r^4)$ $= \tfrac{1}{4}A(R^2 + r^2)$ $= 0.05(D^4 - d^4)$ (approx)	$\dfrac{I}{c} = \dfrac{\pi}{32}\dfrac{D^4 - d^4}{D}$ $= \dfrac{\pi}{4}\dfrac{R^4 - r^4}{R}$ $= 0.8 d_m^2 s$ (approx) when $\dfrac{s}{d_m}$ is very small	$\dfrac{\sqrt{R^2 + r^2}}{2} = \dfrac{\sqrt{D^2 + d^2}}{4}$

17

TABLE 2.1 Properties of Various Cross Sections (*Continued*)

Section	Moment of inertia	Section modulus	Radius of gyration
	$I = r^4 \left(\dfrac{\pi}{8} - \dfrac{8}{9\pi} \right)$ $= 0.1098 r^4$	$\dfrac{I}{c_2} = 0.1908 r^3$ $\dfrac{I}{c_1} = 0.2587 r^3$ $c_1 = 0.4244 r$	$\sqrt{\dfrac{9\pi^2 - 64}{6\pi}}\, r = 0.264 r$
	$I = 0.1098(R^4 - r^4)$ $- \dfrac{0.283 R^2 r^2 (R - r)}{R + r}$ $= 0.3 t r_1^3 \text{ (approx)}$ when $\dfrac{t}{r_1}$ is very small	$c_1 = \dfrac{4}{3\pi} \dfrac{R^2 + Rr + r^2}{R + r}$ $c_2 = R - c_1$	$\sqrt{\dfrac{2I}{\pi(R^2 - r^2)}}$ $= 0.31 r_1 \text{ (approx)}$

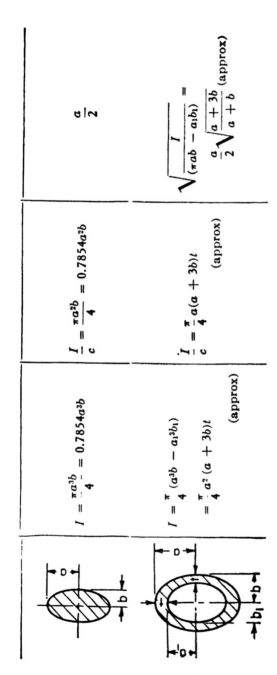

$$I = \frac{\pi a^3 b}{4} = 0.7854 a^3 b$$

$$\frac{I}{c} = \frac{\pi a^2 b}{4} = 0.7854 a^2 b$$

$$\frac{a}{2}$$

$$I = \frac{\pi}{4}(a^3 b - a_1^3 b_1)$$
$$= \frac{\pi}{4} a^2 (a + 3b)t \quad \text{(approx)}$$

$$\frac{I}{c} = \frac{\pi}{4} a(a + 3b)t \quad \text{(approx)}$$

$$\sqrt{\frac{I}{(\pi ab - a_1 b_1)}} =$$
$$\frac{a}{2}\sqrt{\frac{a + 3b}{a + b}} \quad \text{(approx)}$$

19

TABLE 2.1 Properties of Various Cross Sections (*Continued*)

Section	Moment of inertia and section modulus	Radius of gyration
	$$I = \frac{1}{12}\left[\frac{3\pi}{16}d^4 + b(h^3 - d^3) + b^3(h - d)\right]$$ $$\frac{I}{c} = \frac{1}{6h}\left[\frac{3\pi}{16}d^4 + b(h^3 + d^3) + b^3(h - d)\right]$$	$$\sqrt{\frac{I}{\dfrac{\pi d^2}{4} + 2b(h - d)}}$$ (approx)
	$$I = \frac{t}{4}\left(\frac{\pi B^3}{16} + B^2 h + \frac{\pi B h^2}{2} + \frac{2}{3}h^3\right)$$ $$h = H - \tfrac{1}{2}B$$ $$\frac{I}{c} = \frac{2I}{H + t}$$	$$\sqrt{\frac{I}{2\left(\dfrac{\pi B}{4} + h\right)t}}$$

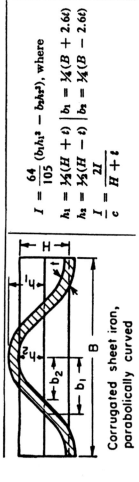

Corrugated sheet iron,
parabolically curved

$I = \dfrac{64}{105}(b_1 h_1^3 - b_2 h_2^3)$, where

$h_1 = \frac{1}{2}(H + t)$ $\quad b_1 = \frac{1}{4}(B + 2.6t)$

$h_2 = \frac{1}{2}(H - t)$ $\quad b_2 = \frac{1}{4}(B - 2.6t)$

$\dfrac{I}{c} = \dfrac{2I}{H + t}$

$r = \sqrt{\dfrac{3I}{t(2B + 5.2H)}}$

TABLE 2.1 Properties of Various Cross Sections (*Continued*)

Approximate values of *least* radius of gyration *r*

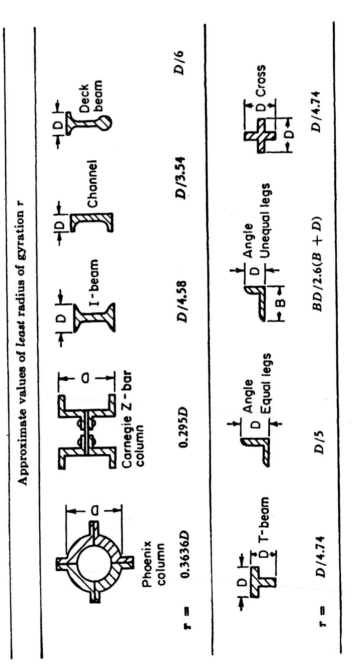

	Phoenix column	Carnegie Z - bar column	I - beam	Channel	Deck beam
r =	0.3636*D*	0.295*D*	*D*/4.58	*D*/3.54	*D*/6

	T-beam	Angle Equal legs	Angle Unequal legs	Cross
r =	*D*/4.74	*D*/5	$BD/2.6(B + D)$	*D*/4.74

22

$$I = I_{cg} + a^2m \qquad \text{for solid body} \qquad (2.6)$$

and
$$I = I_{cg} + a^2A \qquad \text{for plane area} \qquad (2.7)$$

where I_{cg} = moment of inertia of solid, lb·ft² (kg·m²), or area, ft⁴ (m⁴), about axis parallel to reference axis and passing through center of gravity

a = distance between reference axis and axis passing through center of gravity, ft (m)

Likewise,
$$K^2 = K_{cg}^2 + a^2 \qquad (2.8)$$

where K_{cg} = radius of gyration through center of gravity, ft (m).

Polar Moment of Inertia. The polar moment of inertia J of an area is taken about an axis perpendicular to the area and is equal to

$$J = I_1 + I_2 \qquad \text{ft}^4 \text{ (m}^4) \qquad (2.9)$$

where I_1 and I_2 are moments of inertia about any two mutually perpendicular axes lying in the plane of the area and with the intersecting axis perpendicular to the plane.

KINETICS

Energy of a Rigid Body

The *kinetic energy* of a rigid body is the energy possessed by the body by virtue of its motion.

$$\text{Kinetic energy} = \tfrac{1}{2}mv^2 \qquad \text{translation} \qquad (2.10)$$

$$\text{Kinetic energy} = \tfrac{1}{2}I_0\omega^2 \qquad \text{rotation} \qquad (2.11)$$

where m = mass, lb (kg)

I_0 = moment of inertia about axis of rotation, lb·ft² (kg·m²)

v = velocity, ft/s (m/s)

ω = angular velocity, rad/s

The *potential energy* of a rigid body is the energy possessed by the body by virtue of its position, i.e., that energy which is available to do work.

Free Harmonic Vibrations of Systems with 1 Degree of Freedom

If an elastic system is disturbed from its position of equilibrium by a force, the elastic restoring forces of the system in the disturbed position will no longer be in equilibrium with the loading, and vibrations will ensue (Fig. 2.1).

Nomenclature

k = spring constant of elastic system, lbf/ft (N/m)

$v = dx/dt$ = velocity, ft/s (m/s)

v_0 = initial velocity, ft/s (m/s)

t = time, s

W = weight (neglecting spring weight as small compared with weight W), lb (kg)

f = frequency of oscillation, s^{-1}

p = period of oscillation = $\sqrt{kg_c/W}$, s^{-1}

t = time for one complete oscillaton, s

$\omega = p$, in the case of rotation, rad/s

g_c = 32.2 (lbm)(ft)/(lbf)(s^2) (9.81 m/s^2)

x = displacement of W from equilibrium position, ft (m)

x_0 = initial displacement of W from equilibrium position, ft (m)

$$\frac{W}{g_c}\frac{d^2x}{dt^2} - kx = 0 \tag{2.12}$$

$$t = \frac{2\pi}{p} \qquad f = \frac{1}{t} = \frac{p}{2\pi} \qquad p = 2\pi f \qquad p = \frac{2\pi}{t} \tag{2.13}$$

The equation of motion is

$$x = x_0 \cos pt + \frac{v_0}{p} \sin pt \tag{2.14}$$

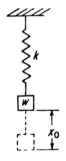

FIGURE 2.1

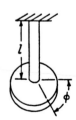

FIGURE 2.2

Natural Frequency. If ∂_{ST} is the deflection of the spring caused by the weight W, then $\partial_{ST} = W/k$ and

$$\omega_n = \sqrt{\frac{g_c}{\partial_{ST}}} = \text{number of free oscillations per } 2\pi \text{ s} \qquad (2.15)$$

Thus the natural frequency is

$$f_n = 3.14 \sqrt{\frac{1}{\partial_{ST}}} \qquad \text{Hz} \qquad (2.16)$$

Torsional Vibration. If a disk is supported as shown in Fig. 2.2 and subjected to a couple in the plane of the disk which is suddenly removed, free torsional vibrations of the elastic system consisting of the shaft and disk will be produced.
Let

ϕ = angle of twist of shaft at any moment, rad
k = torque moment necessary to produce angle of twist of 1 rad in shaft, lbf · ft (N · m)
ω_0 = initial angular velocity, rad/s
ϕ_0 = initial angle of twist of shaft, rad
J = polar moment of inertia of disk (neglecting shaft J as small compared to J of disk), lbm · ft (N · m)

and p, f, t, and g_c are as defined above for elastic vibration. The period of the torsional vibration is

$$p = \sqrt{\frac{k}{J}} \qquad (2.17)$$

The frequency is

$$f = \frac{1}{2\pi} \sqrt{\frac{kg_c}{J}} \qquad (2.18)$$

The equation of motion is

$$\phi = \phi_0 \cos pt + \frac{\omega_0}{p} \sin pt \qquad (2.19)$$

Damped Free Vibrations. Assuming viscous damping, i.e., damping proportional to velocity, such as many exist in dashpots (Fig. 2.3),

$$x = Ae^{(-\alpha+\beta)t} + Be^{(-\alpha-\beta)t} \qquad (2.20)$$

where $\alpha = cg_c/2W$ $\qquad (2.21)$
$\qquad \beta = \sqrt{c^2g_c^2/4W^2 - kg_c/W}$ $\qquad (2.22)$
$\qquad A, B$ = constants of integration
$\qquad c$ = damping coefficient, lbf/ft (N/m)

When $c^2g_c^2/4W^2 > kg_c/W$, β is real and positive. The result is exponential decay, in which $x \to 0$ and $t \to \infty$.

When $c^2g_c^2/4W^2 < kg_c/W$, β is imaginary. This case is more representative of the usual case of damped vibration, with the amplitude diminishing after each cycle according to the physical constants of the system. The frequency, however, does not change; that is, $t_1 = t_2 = t_3$ in Fig. 2.4.

When $c^2g_c^2/4W^2 = kg_c/W$, $\beta = 0$. For this condition

$$c = c_{cr} = \sqrt{\frac{4Wk}{g_c}} \qquad (2.23)$$

or critical damping (see Fig. 2.5). This is a boundary case and rarely exists.

Forced Vibrations without Damping

P = impressed force, lbf, with frequency ω, s^{-1}
$\quad$ = $P_0 \cos \omega t$, where t is time for one vibration, s
p = natural frequency of system, s^{-1}, of weight W, lbm (kg)
g_c = 32.2 (lbm · ft)/(lbf · s^2) (9.81 m/s^2)

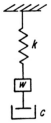

FIGURE 2.3

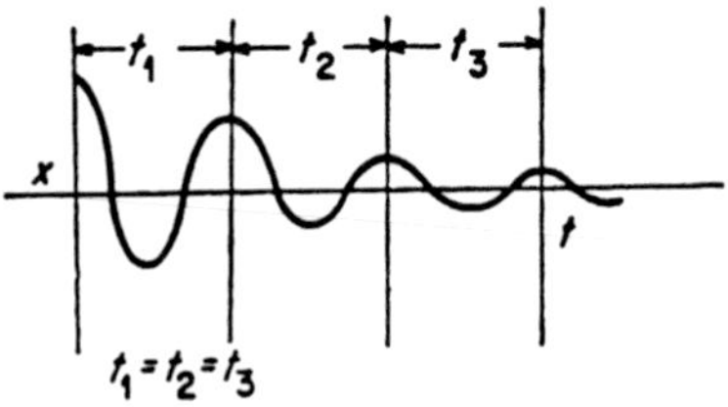

FIGURE 2.4

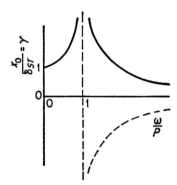

FIGURE 2.5   **FIGURE 2.6**

$$x = A \sin pt + B \cos pt + \frac{P_0 g_c}{W} \frac{1}{p^2 - \omega^2} \cos \omega t \qquad (2.24)$$

where A and B are constants of integration.

In general, the vibrations due to the first two terms die out shortly, and only those remain that are due to the forcing frequency ω. At such time, then,

$$x = \frac{P_0 g_c}{W} \frac{1}{p^2 - \omega^2} \cos \omega t \qquad (2.25)$$

Let $\partial_{ST} = P_0/k =$ static deflection, in feet meters (m), resulting from P_0, where k is the system constant, in pounds force per foot (N/m) and

$$x_0 = x_{max} = \frac{P_0 g_c}{W} \frac{1}{p^2 - \omega^2} \qquad (2.26)$$

Then
$$\frac{x_0}{\partial_{ST}} = \frac{1}{1 - \omega^2/p^2} = \gamma \qquad (2.27)$$

This relation can be plotted as shown in Fig. 2.6. It is seen that

1. When $\omega/p = 0$, or when ω is small compared to p, $x = \delta_{ST}$ or nearly so.
2. When $\omega/p = \infty$, that is, when ω is very large compared to p, $x = 0$.
3. When $\omega/p = 1$, or $\omega = p$, $x = \infty$. This is the case of resonance.

Figure 2.6 shows that when the applied force frequency becomes larger than the natural frequency of the body, the deflection of the body is opposite in direction to that of the force.

In vibration isolation, a quantity known as the *transmission ratio* is defined as equal to $1/(\omega^2/p^2 - 1)$. For practical vibration isolation, $\omega/p \geq \sqrt{2}$. This is accomplished by supplying a small value of p through use of very soft springs or by increasing the mass of the machine or its foundation.

Forced Vibrations with Damping (Fig. 2.7)

The motion of the weight W at any time is

$$x = e^{-\alpha t}(A \cos \beta t + B \sin \beta t) + C \sin \omega t + D \cos \omega t \quad (2.28)$$

where A, B, C, and D are constants of integration and α and β are as defined by Eqs. (2.21) and (2.22).

For steady-state application, only the last two terms are of interest, that is,

$$x = C \sin \omega t + D \cos \omega t \quad (2.29)$$

where

$$C = \frac{P_0 g_c}{W} \frac{W^2 g_c}{(p^2 - \omega^2) W^2 + c^2 g_c^2 \omega^2} \quad (2.30)$$

$$D = \frac{P_0 g_c}{W} \frac{W^2(p^2 - \omega^2)}{(p^2 - \omega^2)^2 W^2 + c^2 g_c^2 \omega^2} \quad (2.31)$$

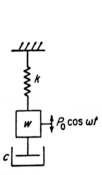

FIGURE 2.7

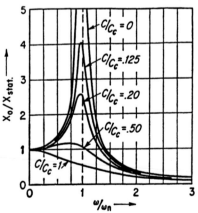

FIGURE 2.8

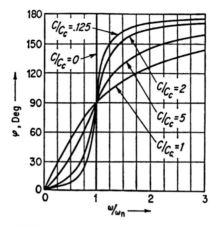

FIGURE 2.9

Let

$$R = \sqrt{c^2 + d^2} = \frac{P_0}{k} \frac{1}{\sqrt{(1 - \omega^2/p^2)^2 + (2c\omega/C_{cr}p)^2}} \qquad (2.32)$$

and

$$\theta = \tan^{-1} \frac{C}{D} = \tan^{-1} \frac{cg_c\omega}{W(p^2 - \omega^2)} \qquad (2.33)$$

Figures 2.8 and 2.9 can be drawn as above.

TORSION

See Table 2.2.

Torsion (Solid Circular Shafts)

See Fig. 2.10.

$$S_v = \frac{M_t c}{J} \qquad (2.34)$$

where S_v = shear stress, psi (MPa)
M_t = twisting moment = Pl, in·lb (N·m)
c = distance from center to stressed surface of interest, in (mm)
J = polar moment of inertia of cross section, in⁴ (mm⁴)

TABLE 2.2 Torsion of Shafts of Various Cross Sections

G = Shear Modulus of Elasticity, psi

Cross section	Torsional resisting moment M_t	Angular deflection, a_1 (length = 1 in. (25.4 mm), radius = 1 in.)25.4 mm)		Work of torsion (V = volume)
		In terms of torsional moment	In terms of max shear	
(solid circle, diameter d)	$\dfrac{\pi}{16} d^3 S_v$	$\dfrac{M_t}{GJ} = \dfrac{32}{\pi d^4} \dfrac{M_t}{R}$	$\dfrac{S_{v\,max}}{2} \dfrac{1}{G} \dfrac{1}{d}$	$\dfrac{1}{4} \dfrac{S^2_{v\,max}}{G} V$ (Note 1)
(hollow circle, inner d, outer D)	$\dfrac{\pi}{16} \dfrac{D^4 - d^4}{D} S_v$	$\dfrac{32}{\pi(D^4 - d^4)} \dfrac{M_t}{G}$	$\dfrac{S_{v\,max}}{2} \dfrac{1}{G} \dfrac{1}{D}$	$\dfrac{1}{4} \dfrac{S^2_{v\,max}}{G} \dfrac{D^2 + d^2}{D^2} V$ (Note 2)

	$\dfrac{\pi}{16} b^2 h S_v$ $(h > b)$	$\dfrac{16}{\pi} \dfrac{b^2 + h^2}{b^3 h^3} \dfrac{M_t}{G}$	$\dfrac{S_{v\,\max}}{G} \dfrac{b^2 + h^2}{b h^2}$	$\dfrac{1}{8} \dfrac{S^2_{v\,\max}}{G} \dfrac{b^2 + h^2}{h^2} V$ (Note 3)
	$\tfrac{2}{9} b^2 h S_v$ $(h > b)$	$3.6 \dfrac{b^2 + h^2}{b^3 h^3} \dfrac{M_t}{G}$ *	$0.8 \dfrac{S_{v\,\max}}{G} \dfrac{b^2 + h^2}{b h^2}$ *	$\dfrac{4}{45} \dfrac{S^2_{v\,\max}}{G} \dfrac{b^2 + h^2}{h^2} V$ (Note 4)

TABLE 2.2 Torsion of Shafts of Various Cross Sections (*Continued*)

Cross section	Torsional resisting moment M_t	Angular deflection, a_1 (length = 1 in. (25.4 mm), radius = 1 in.)25.4 mm))		Work of torsion (V = volume)
		In terms of torsional moment	In terms of max shear	
	$\frac{2}{9}h^3 S_v$	$7.2 \frac{1}{h^4}\frac{M_t}{G}$	$1.6 \frac{S_{v_{max}}}{G}\frac{1}{h}$	$\frac{8}{45}\frac{S^2{v_{max}}}{G}V$ (Note 5)

	$\dfrac{b^3}{20} S_w$	$4.62 \dfrac{1}{b^4} \dfrac{M_t}{G}$	$2.31 \dfrac{S_{w_{max}}}{G} \dfrac{1}{b}$
	$\dfrac{b^3}{1.09} S_w$	$0.967 \dfrac{1}{b^4} \dfrac{M_t}{G}$	$0.9 \dfrac{S_{w_{max}}}{G} \dfrac{1}{b}$

* When

Coefficient 3.6 becomes = 3.56
Coefficient 0.8 becomes = 0.79

$h/b =$	1	2	4	8
	3.56	3.50	3.35	3.21
	0.79	0.78	0.74	0.71

NOTES: (1) $S_{w_{max}}$ at circumference. (2) $S_{w_{max}}$ at outer circumference. (3) $S_{w_{max}}$ at A; $S_{w_B} = 16M_t/\pi b h^2$.
(4) $S_{w_{max}}$ at middle of side h; in middle of b, $S_w = 9M_t/2bh^2$. (5) $S_{w_{max}}$ at middle of side.

33

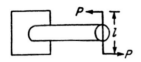

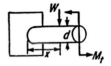

FIGURE 2.10   **FIGURE 2.11**

Combined Torsion and Bending (Solid Circular Shafts)

See Fig. 2.11.

$$\sigma_{max} = \frac{16}{\pi d^3}(M_b + \sqrt{M_{b^2} + M_{t^2}}) \qquad (2.35)$$

where σ_{max} = maximum stress, psi (MPa)
M_t = torque, in · lb (N · m)
M_b = moment due to bending load, in · lb (N · m) = W_x
d = diameter of bar, in (mm)

and

$$M = \frac{\sigma I}{c} \qquad (2.36)$$

where M = bending moment, lb · in (N · m)
σ = elastic stress at distance c from neutral axis, psi (MPa)
c = distance from neutral axis to plane at which stress σ is calculated, in (mm)
I = rectangular moment of inertia of cross-sectional area about neutral axis, in⁴ (mm⁴)
I/c = section modulus where c is distance to outermost fiber, in³ (mm³)

CYLINDER STRESSES

Stresses in Thin-Walled Tubes or Cylinders

See Fig. 2.11.

$$\sigma_h = \frac{pd}{2t} \qquad \sigma_l = \frac{pd}{4t} \qquad (2.37)$$

where σ_h = hoop stress, psi (MPa)
σ_l = longitudinal stress, psi (MPa)

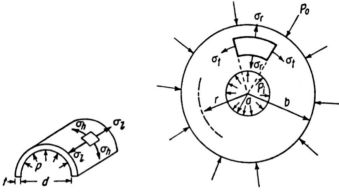

FIGURE 2.12   **FIGURE 2.13**

d = internal diameter, in (mm)
p = internal pressure, psi (MPa)
t = thickness of tube wall, in (mm)

Stresses in Thick Cylinders or Tubes (Figs. 2.12 and 2.13)

For internal pressure only:

$$\sigma_r = \frac{a^2 p_i}{b^2 - a^2}\left(1 - \frac{b^2}{r^2}\right) \tag{2.38}$$

$$\sigma_t = \frac{a^2 p_i}{b^2 - a^2}\left(1 + \frac{b^2}{r^2}\right) \tag{2.39}$$

For external pressure only:

$$\sigma_r = -\frac{p_0 b^2}{b^2 - a^2}\left(1 - \frac{a^2}{r^2}\right) \tag{2.40}$$

$$\sigma_t = -\frac{p_0 b^2}{b^2 - a^2}\left(1 + \frac{a^2}{r^2}\right) \tag{2.41}$$

where σ_r = stress in radial direction, psi (MPa)
$\quad\sigma_t$ = stress in tangential direction, psi (MPa)
$\quad a$ = internal radius of cylinder, in (mm)
$\quad b$ = external radius of cylinder, in (mm)
$\quad r$ = radial measurement, in (mm)
$\quad p_i$ = internal pressure, psi (MPa)
$\quad p_0$ = external pressure, psi (MPa)

SECTION 3
FORMULAS FOR STRESSES IN MACHINE MEMBERS

NORMAL AND PRINCIPAL STRESSES

Normal Stresses

The maximum and minimum normal stresses, $s_n(\text{max})$ and $s_n(\text{min})$, which are tensile or compressive stresses, can be determined for the general case of two-dimensional loading on a particle by

$$s_n(\text{max}) = \frac{s_x + s_y}{2} + \sqrt{\left(\frac{s_x - s_y}{2}\right)^2 + \tau_{xy}^2} \qquad (3.1)$$

$$s_n(\text{min}) = \frac{s_x + s_y}{2} - \sqrt{\left(\frac{s_x - s_y}{2}\right)^2 + \tau_{xy}^2} \qquad (3.2)$$

Equations (3.1) and (3.2) give algebraic maximum and minimum values, where

s_x is a stress at a critical point in tension or compression normal to the cross section under consideration, it may be due to either bending or axial loads, or to a combination of the two. When s_x is in tension, it must be preceded by a plus $(+)$ sign, and when it is in compression, it must be preceded by a minus $(-)$ sign.

s_y is a stress at the same critical point and in a direction normal to the s_x stress. Again, this stress must be preceded by the proper algebraic sign.

τ_{xy} is the *shear stress* at the same critical point acting in the plane normal to the y axis (which is the xz plane) and in the plane normal to the x axis (which is the yz plane). This shear stress may be due to a torsional moment, a transverse load, or to a combination of the two. The manner in which these stresses are oriented with respect to each other is shown in Fig. 3.1*a*.

s_n(max) and s_n(min) are called *principal stresses* and occur on planes that are at 90° to each other, called *principal planes*. These are also planes of zero shear. For two-dimensional loading, the third principal stress is zero. The manner in which the principal stresses are oriented with respect to each other is shown in Fig. 3.1*b*.

Maximum Shear Stress

The maximum shear stress τ(max) at the critical point being investigated is equal to one-half of the greatest difference of any two of the three principal stresses (do not overlook any of the principal stresses which are zero). Hence, for the case of two-dimensional loading on a particle causing a two-dimensional stress,

$$\tau(\text{max}) = \frac{s_n(\text{max}) - s_n(\text{min})}{2} \quad \text{or} \quad \frac{s_n(\text{max}) - 0}{2}$$

$$\text{or} \quad \frac{s_n(\text{min}) - 0}{3} \tag{3.3}$$

depending upon which results in the greatest numerical value. The planes of maximum shear are inclined at 45° with the principal planes.

Application

Application of Eqs. (3.1) and (3.2) requires the determination of s_x, s_y, and τ_{xy} at the critical point in the machine member. The critical point is the point at which the applied loads produce the maximum combined stress

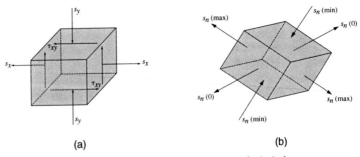

(a)   (b)

FIGURE 3.1 (*a*) Shear-stress orientation; (*b*) Orientation of principal stresses.

effects. In a beam, the following are representative stresses that can occur, to be included in Eqs. (3.1) and (3.2) if they act at the same point.

$$s_x \quad \text{and} \quad s_y = \pm \frac{Mc}{l} \pm \frac{P}{A} \qquad (3.4)$$

remembering that these stresses may be either plus or minus depending upon whether they are tension or compression.

$$\tau_{xy} = \frac{Tr}{J} + s_v \qquad (3.5)$$

for a circular cross section (when these stresses are parallel)

where M = bending moment, in · lb (N · m)
c = distance from neutral axis to outer surface, in (mm)
r = radius of circular cross section, in (mm)
I = rectangular moment of inertia of cross section, in⁴ (mm⁴)
P = axial load, lb (N)
A = area of cross section, in² (mm²)
T = torsional moment, in · lb (N · m)
J = polar moment of inertia of cross section, (mm⁴)
s_v = transverse shear, psi (MPa)

$$s_v = \frac{VQ}{lb} \qquad (3.6)$$

where V = transverse shear load on cross section, lb (N)
b = width of section containing critical point, in (mm)
Q = moment of cross-sectional area of member, above or below critical point, with respect to neutral axis, in³ (mm³)

$s_v(\text{max}) = \dfrac{4V}{3A}$ for circular cross section and occurs at

neutral axis (3.7)

$s_v(\text{max}) = \dfrac{3V}{2A}$ for rectangular cross section and

occurs at neutral axis (3.8)

$s_n(\text{max})$ = maximum algebraic stress, psi (MPa)

$s_n(\text{min})$ = minimum algebraic stress, psi (MPa)

$\tau(\text{max})$ = maximum shear stress, psi (MPa)

STRESSES DUE TO INTERFERENCE FITS

These may be calculated by considering the fitted parts as thick-walled cylinders, as shown in Fig. 3.2, by the following equations:

$$P_C = \frac{\delta}{d_C \left[\dfrac{d_c^2 + d_i^2}{E_i(d_c^2 - d_i^2)} + \dfrac{d_o^2 + d_c^2}{E_o(d_o^2 - d_c^2)} - \dfrac{\mu_i}{E_i} + \dfrac{\mu_{;o}}{E_o} \right]} \qquad (3.9)$$

where p_c = pressure at contact surface, psi (MPa)
δ = total interference, in (mm)
d_i = inside diameter of inner member, in (mm)
d_c = diameter of contact surface, in (mm)
d_o = outside diameter of outer member, in (mm)
μ_o = Poisson's ratio for outer member
μ_i = Poisson's ratio for inner member
E_o = modulus of elasticity of outer member, psi (MPa)
E_i = modulus of elasticity of inner member, psi (MPa)

If the outer and inner members are of the same material, Equation (3.9) reduces to

$$p_c = \frac{\delta}{2d_c^3(d_o^2 - d_i^2)/[E(d_c^2 - d_i^2)(d_o^2 - d_c^2)]} \qquad (3.10)$$

After p_c has been determined, then the actual tangential stresses at the various surfaces, in accordance with Lamé's equation, for use in conjunction with the maximum shear theory of failure, may be determined by

On the surface at d_o,

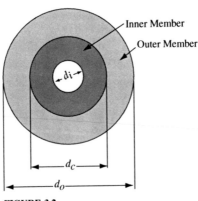

Inner Member
Outer Member

d_i
d_c
d_o

FIGURE 3.2

$$s_{to} = \frac{2p_c d_c^2}{d_o^2 - d_c^2} \qquad (3.11)$$

On the surface at d_c for the outer member,

$$s_{tco} = p_c \left(\frac{d_o^2 + d_c^2}{d_o^2 - d_c^2} \right) \qquad (3.12)$$

On the surface at d_c for the inner member,

$$s_{tci} = -p_c \left(\frac{d_c^2 + d_i^2}{d_c^2 - d_i^2} \right) \qquad (3.13)$$

On the surface at d_i,

$$s_{ti} = \frac{-2p_c d_c^2}{d_c^2 - d_i^2} \qquad (3.14)$$

The equivalent tangential stresses at the various surfaces, in accordance with Birnie's equation, for use in conjunction with the maximum strain theory of failure may be determined by

On the surface at d_o for the outer member,

$$s'_{to} = \frac{2p_c d_c^2}{d_o^2 - d_c^2} \qquad (3.15)$$

On the surface at d_c for the outer member,

$$s'_{tco} = p_c \left(\frac{d_o^2 + d_c^2}{d_o^2 - d_c^2} + \mu_o \right) \qquad (3.16)$$

On the surface at d_c for the inner member,

$$s'_{tci} = - p_c \left(\frac{d_c^2 + d_i^2}{d_c^2 - d_i^2} - \mu_i \right) \qquad (3.17)$$

On the surface at d_i,

$$s'_{ti} = \frac{-2p_c d_c^2}{d_c^2 - d_i^2} \qquad (3.18)$$

Forces and Torques

The maximum axial force F_a required to assemble a force fit varies directly as the thickness of the outer member, the length of the outer member, the difference in diameters of the mating members, and the coefficient of friction. This force in pounds may be approximated by

$$F_\alpha = f\pi d L p_c \qquad (3.19)$$

The torque that can be transmitted by an interference fit without slipping between the hub and shaft can be estimated by

$$T = \frac{f p_c \pi d^2 L}{2} \qquad (3.20)$$

where F_α = axial load, lb (N)
T = torque transmitted, in · lb (N · m)
d = nominal shaft diameter, in (mm)
f = coefficient of friction
L = length of external member, in (mm)
p_c = contact pressure between two members, psi (MPa)

Assembly of Shrink-Fits

This is often facilitated by heating the hub until it has expanded by an amount at least as much as the interference. The temperature change ΔT required to effect an increase δ in the inside diameter of the hub may be determined by

$$\Delta T = \frac{\delta}{\alpha d_i} \qquad (3.21)$$

where δ = diametral interference, in (mm)
α = coefficient of linear expansion, per °F (°C)
ΔT = change in temperature, °F (°C)
d_i = initial diameter of hole before expansion, in (mm)

An alternate to heating the hub is to cool the shaft by means of a coolant such as dry ice.

BEAM FORMULAS

See Table 3.1 and Fig. 3.3.

TABLE 3.1 Beam Formulas

DIAGRAMS	REACTIONS=R SHEAR=V	BENDING MOMENT = M	DEFLECTION = D
	CASE I. - Beam Supported Both Ends - Continuous Load, Uniformly Distributed.		
	$$R = R_1 = V\,(\text{max.}) = \frac{W}{2}$$ At x: $$V = \frac{W}{2} - \frac{Wx}{L}$$	At center: $$M\,(\text{max.}) = \frac{WL}{8}$$ At x: $$M = \frac{Wx}{2L}\,(L-x)$$	At center: $$D\,(\text{max.}) = \frac{5}{384}\frac{WL^3}{EI}$$ At x: $$D = \frac{Wx}{24\,EIL}\,(L^3 - 2Lx^2 + x^3)$$
	CASE 2. - Beam Supported Both Ends - Concentrated Load at Any Point.		
	$$R = \frac{Wb}{L}$$ $$R_1 = \frac{Wa}{L}$$ $V\,(\text{max.}) = R$ when $a < b$ and R_1 when $a > b$ At x: $$V = \frac{Wb}{L}$$	At point of load: $$M\,(\text{max.}) = \frac{Wab}{L}$$ At x: when $x < a$ $$M = \frac{Wbx}{L}$$	At x: when $x = \sqrt{a(a+2b)} \div 3$ and $a > b$ $$D\,(\text{max.}) = Wab\,(a+2b)\sqrt{3a(a+2b)} \div 27\,EIL$$ At x: when $x < a$ $$D = \frac{Wbx}{6\,EIL}\left[2L(L-x) - b^2 - (L-x)^2\right]$$ At x: when $x > a$ $$D = \frac{Wa(L-x)}{6\,EIL}\left[2Lb - b^2 - (L-x)^2\right]$$

43

TABLE 3.1 Beam Formulas (*Continued*)

CASE 3. - Beam Supported Both Ends - Two Unequal Concentrated Loads, Unequally Distributed.

$$R = \frac{1}{L}\left[W_o(L-a) + W_1 b\right]$$

$$R_1 = \frac{1}{L}\left[W_o a + W_1(L-b)\right]$$

V (max.) = Maximum Reaction

At x when $x > a$ and $< (L-b)$
$$V = R - W$$

At point of load W:
$$M = \frac{a}{L}\left[W_o(L-a) + W_1 b\right]$$

At point of load W_1:
$$M_1 = \frac{b}{L}\left[W_o a + W_1(L-b)\right]$$

At x when $x > a$ or $< (L-b)$
$$M = W_o \frac{a}{L}(L-x) + W_1\frac{bx}{L}$$

CASE 4. - Beam Supported Both Ends - Three Unequal Concentrated Loads, Unequally Distributed.

$$R = \frac{W b + W_1 b_1 + W_2 b_2}{L}$$

$$R_1 = \frac{W_o a + W_1 a_1 + W_2 a_2}{L}$$

V (max.) = Maximum Reaction

At x when $x > a$ and $< a_1$
$$V = R - W$$

At x when $x > a_1$ and $< a_2$
$$V = R - W - W_1$$

At x when $x = a$
$$M = Ra$$

At x when $x = a_1$
$$M_1 = Ra_1 - W(a_1 - a)$$

At x when $x = a_2$
$$M_2 = Ra_2 - W(a_2 - a) - W_1(a_2 - a_1)$$

M (max.) = M when W = R or > R

M (max.) = M_1 when $\begin{cases} W_1 + W = R \text{ or } > R \\ W_1 + W_2 = R_1 \text{ or } > R_1 \end{cases}$

M (max.) = M_2 when $W_2 = R_1$ or $> R_1$

44

CASE 5. - Beam Fixed Both Ends - Continuous Load, Uniformly Distributed.

$$R = R_1 = V \text{ (max.)} = \frac{W}{2}$$

At x:

$$V = \frac{W}{2} - \frac{Wx}{L}$$

At center:

$$M \text{ (max.)} = \frac{WL}{24}$$

At supports:

$$M_1 \text{ (max.)} = \frac{WL}{12}$$

At x:

$$M = \frac{W}{2L}\left(-\frac{L^2}{6} + Lx - x^2\right)$$

At center:

$$D \text{ (max.)} = \frac{1}{384}\frac{WL^3}{EI}$$

At x:

$$D = \frac{Wx^2}{24 EIL}\left(l^2 - 2Lx + x^2\right)$$

CASE 6. - Beam Fixed Both Ends - Concentrated Load at Any Point.

$$R = W\left(\frac{b^2(3a+b)}{L^3}\right)$$

$$R_1 = W\left(\frac{a^2(3b+a)}{L^3}\right)$$

$$V \text{ (max.)} = R \text{ when } a < b$$
$$= R_1 \text{ when } a > b$$

At x when x < a

$$V = R$$

At support R:

$$M_1\left(\begin{array}{c}\text{max. neg. mom.}\\ \text{when } b > a\end{array}\right) = -W\,\frac{ab^2}{L^2}$$

At support R_1:

$$M_2\left(\begin{array}{c}\text{max. neg. mom.}\\ \text{when } a > b\end{array}\right) = -W\,\frac{a^2b}{L^2}$$

At point of load:

$$M \text{ (max.)} = R_a + M_1 = R_a - W\,\frac{ab^2}{L^2}$$

At x: $M = R_x - W\,\frac{ab^2}{L^2}$

At x when $x = \frac{2aL}{3a+b}$ and a > b

$$D \text{ (max.)} = \frac{2W a^3 b^2}{3 EI (3a+b)^2}$$

when x < a

$$D = \frac{W\,b^2x^2}{6\,EIL^3}\,(3aL - 3ax - bx)$$

45

TABLE 3.1 Beam Formulas (*Continued*)

DIAGRAMS	REACTIONS=R SHEAR=V	BENDING MOMENT=M	DEFLECTION=D
	CASE 7. - Beam Fixed at One End (Cantilever) - Continuous Load, Uniformly Distributed..		
	$R_1 = V(\text{max.}) = W$ At x: $\quad V = \dfrac{Wx}{L}$	At fixed end: $\quad M(\text{max.}) = \dfrac{WL}{2}$ At x: $\quad M = \dfrac{Wx^2}{2L}$	At free end: $\quad D(\text{max.}) = \dfrac{WL^3}{8EI}$ At x: $\quad D = \dfrac{W}{24\,EIL}\,(x^4 - 4L^3x + 3L^4)$
	CASE 8. - Beam Fixed at One End (Cantilever) - Concentrated Load at Any Point.		
	$R_1 = V(\text{max.}) = W$ At x when $x > a$: $\quad V = W$ At x when $x < a$: $\quad V = 0$	At fixed end: $\quad M(\text{max.}) = Wb$ At x when $x > a$: $\quad M = W(x-a)$	At free end: $\quad D(\text{max.}) = \dfrac{Wb^2}{6EI}\left[2 - \dfrac{3a}{L} + \left(\dfrac{a}{L}\right)^3\right]$ At point of load: $\quad D = \dfrac{W}{3EI}(L-a)^3$ At x when $x > a$: $\quad D = \dfrac{W}{6EI}\left(-3aL^2 + 2L^3 + x^3 - 3ax^2 - 3L^2x + 6aLx\right)$

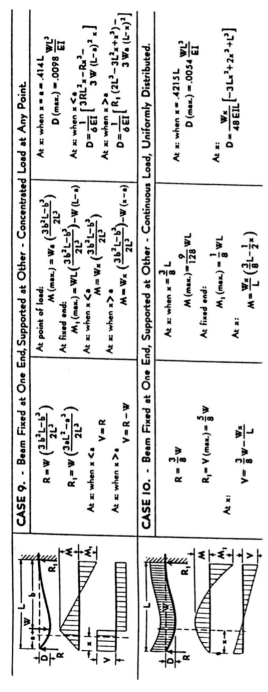

CASE 9. - Beam Fixed at One End, Supported at Other - Concentrated Load at Any Point

$$R = W\left(\frac{3b^2L - b^3}{2L^3}\right)$$

$$R_1 = W\left(\frac{3aL^2 - a^3}{2L^3}\right)$$

At x: when x < a

$$V = R$$

At x: when x > a

$$V = R - W$$

At point of load:

$$M\,(max.) = W_a\left(\frac{3b^2L - b^3}{2L^3}\right)$$

At fixed end:

$$M_1\,(max.) = WL\left(\frac{3b^2L - b^3}{2L^3}\right) - W(L - a)$$

At x: when x < a

$$M = W_x\left(\frac{3b^2L - b^3}{2L^3}\right)$$

At x: when x > a

$$M = W_x\left(\frac{3b^2L - b^3}{2L^3}\right) - W(x - a)$$

At x: when x = a = .414L

$$D\,(max.) = .0098\,\frac{WL^3}{EI}$$

At x: when x < a

$$D = \frac{1}{6EI}\left[\,3RL^2 x - R_a^3 - \frac{3W}{3W}(L - a)^2 x\,\right]$$

At x: when x > a

$$D = \frac{1}{6EI}\left[\,R_1\,(2L^3 - 3L^2 x + x^3) - 3W_a\,(L - x)^2\,\right]$$

CASE 10. - Beam Fixed at One End, Supported at Other - Continuous Load, Uniformly Distributed.

$$R = \frac{3}{8}W$$

$$R_1 = V\,(max.) = \frac{5}{8}W$$

At x:

$$V = \frac{3}{8}W - \frac{W_x}{L}$$

At x: when x = $\frac{3}{8}$L

$$M\,(max.) = \frac{9}{128}WL$$

At fixed end:

$$M_1\,(max.) = \frac{1}{8}WL$$

At x:

$$M = \frac{W_x}{L}\left(\frac{3}{8}L - \frac{1}{2}x\right)$$

At x: when x = .4215L

$$D\,(max.) = .0054\,\frac{WL^3}{EI}$$

At x:

$$D = \frac{W_x}{48\,EIL}\left[-3Lx^2 + 2x^3 + L^3\right]$$

47

TABLE 3.1 Beam Formulas *(Continued)*

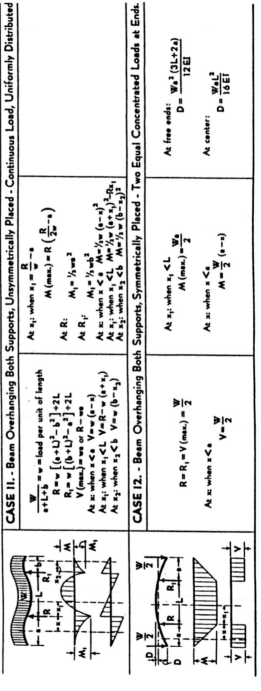

CASE 11. – Beam Overhanging Both Supports, Unsymmetrically Placed - Continuous Load, Uniformly Distributed

$\dfrac{W}{a+L+b} = w = $ load per unit of length

$R = w\left[(a+L)^2 - b^2\right] \div 2L$

$R_1 = w\left[(b+L)^2 - a^2\right] \div 2L$

$V(\text{max.}) = wa$ or $R - wa$

At x: when $x < a$ $V = w(a-x)$

At x_1: when $x_1 < L$ $V = R - w(a+x_1)$

At x_2: when $x_2 < b$ $V = w(b-x_2)$

At x_1: when $x_1 = \dfrac{R}{w} - a$

$M(\text{max.}) = R\left(\dfrac{R}{2w} - a\right)$

At R: $M_1 = \frac{1}{2}wa^2$

At R_1: $M_1 = \frac{1}{2}wb^2$

At x: when $x < a$ $M = \frac{1}{2}w\,(a-x)^2$

At x_1: when $x_1 < L$ $M = \frac{1}{2}w\,(a+x_1)^2 - Rx_1$

At x_2: when $x_2 < b$ $M = \frac{1}{2}w\,(b-x_2)^2$

CASE 12. – Beam Overhanging Both Supports, Symmetrically Placed - Two Equal Concentrated Loads at Ends.

$R = R_1 = V(\text{max.}) = \dfrac{W}{2}$

At x: when $x < a$

$V = \dfrac{W}{2}$

At x_1: when $x_1 < L$

$M(\text{max.}) = \dfrac{Wa}{2}$

At x: when $x < a$

$M = \dfrac{W}{2}(a-x)$

At free ends: $D = \dfrac{Wa^2(3L+2a)}{12EI}$

At center: $D = \dfrac{WaL^2}{16EI}$

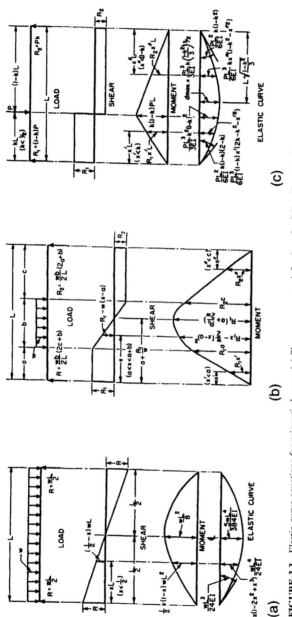

FIGURE 3.3 Elastic-curve equations for prismatic beams. (*a*) Shears, moments, deflections for full uniform load on a simply supported prismatic beam. (*b*) Shears and moments for uniform load over part of a simply supported prismatic beam. (*c*) Shears, moments, deflections for a concentrated load at any point of a simply supported prismatic beam.

49

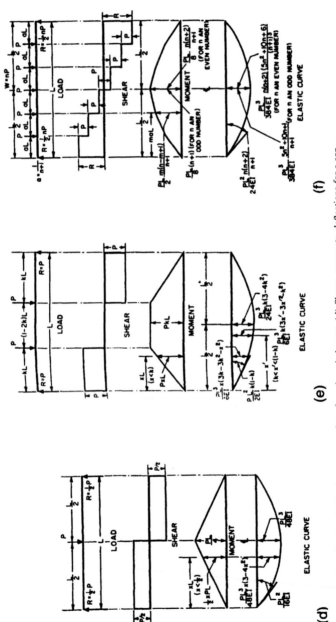

FIGURE 3.3 (*Continued*) Elastic-curve equations for prismatic beams. (*d*) Shears, moments, deflections for a concentrated load at midspan of a simply supported prismatic beam. (*e*) Shears, moments, deflections for two equal concentrated loads on a simply supported prismatic beam. (*f*) Shears, moments, deflections for several equal loads equally spaced on a simply supported prismatic beam.

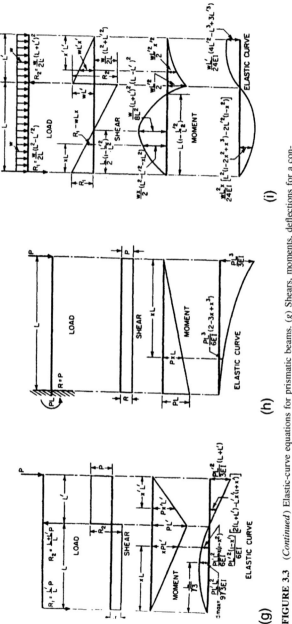

FIGURE 3.3 (*Continued*) Elastic-curve equations for prismatic beams. (*g*) Shears, moments, deflections for a concentrated load on a beam overhang. (*h*) Shears, moments, deflections for a concentrated load on the end of a prismatic cantilever. (*i*) Shears, moments, deflections for a uniform load over the full length of a beam with overhang.

51

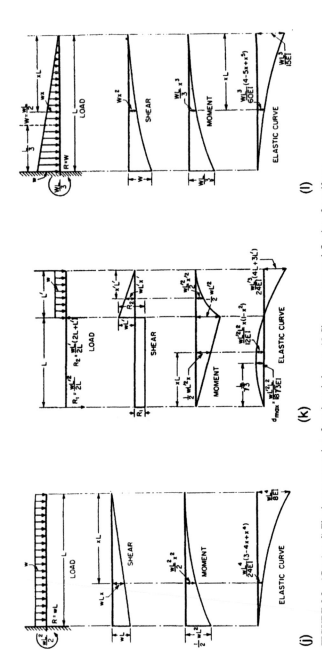

FIGURE 3.3 (*Continued*) Elastic-curve equations for prismatic beams. (*j*) Shears, moments, deflections for uniform load over the full length of a cantilever. (*k*) Shears, moments, deflections for uniform load on a beam overhang. (*l*) Shears, moments, deflections for triangular loading on a prismatic cantilever.

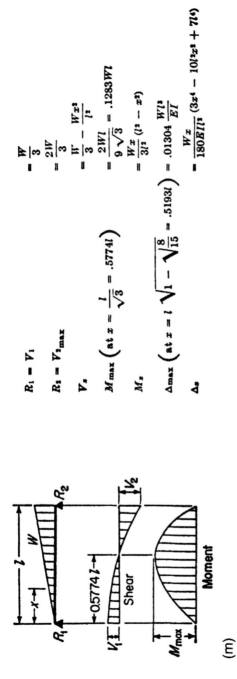

$$R_1 = V_1 = \frac{W}{3}$$

$$R_2 = V_{2\,\text{max}} = \frac{2W}{3}$$

$$V_x = \frac{W}{3} - \frac{Wx^2}{l^2}$$

$$M_{\text{max}}\left(\text{at } x = \frac{l}{\sqrt{3}} = .5774l\right) = \frac{2Wl}{9\sqrt{3}} = .1283Wl$$

$$M_x = \frac{Wx}{3l^2}(l^2 - x^2)$$

$$\Delta_{\text{max}}\left(\text{at } x = l\sqrt{1 - \sqrt{\frac{8}{15}}} = .5193l\right) = .01304\,\frac{Wl^3}{EI}$$

$$\Delta_x = \frac{Wx}{180EIl^2}(3x^4 - 10l^2x^2 + 7l^4)$$

FIGURE 3.3 (*Continued*) Elastic-curve equations for prismatic beams. (*m*) Simple beam—load increasing uniformly to one end.

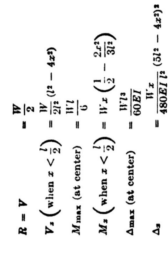

$$R = V = \frac{W}{2}$$

$$V_x \left(\text{when } x < \frac{l}{2} \right) = \frac{W}{2l^2}(l^2 - 4x^2)$$

$$M_{max} \text{ (at center)} = \frac{Wl}{6}$$

$$M_x \left(\text{when } x < \frac{l}{2} \right) = Wx \left(\frac{1}{2} - \frac{2x^2}{3l^2} \right)$$

$$\Delta_{max} \text{ (at center)} = \frac{Wl^3}{60EI}$$

$$\Delta_x = \frac{Wx}{480EIl^2}(5l^2 - 4x^2)^2$$

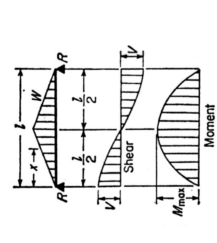

(n)

FIGURE 3.3 (*Continued*) Elastic-curve equations for prismatic beams. (*n*) Simple beam—load increasing uniformly to center.

$R_1 = V_{1\max}$ $\quad = \dfrac{wa}{2l}(2l - a)$

$R_2 = V_2$ $\quad = \dfrac{wa^2}{2l}$

V (when $x < a$) $\quad = R_1 - wx$

$M_{\max}\left(\text{at } x = \dfrac{R_1}{w}\right) = \dfrac{R_1{}^2}{2w}$

M_x (when $x < a$) $\quad = R_1 x - \dfrac{wx^2}{2}$

M_x (when $x > a$) $\quad = R_2(l - x)$

Δ_x (when $x < a$) $\quad = \dfrac{wx}{24EIl}[a^2(2l - a)^2 - 2ax^2(2l - a) + lx^3]$

Δ_x (when $x > a$) $\quad = \dfrac{wa^2(l - x)}{24EIl}(4xl - 2x^2 - a^2)$

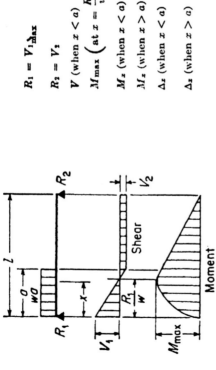

(o)

FIGURE 3.3 (*Continued*) Elastic-curve equations for prismatic beams. (*o*) Simple beam—uniform load partially distributed at one end.

$$R = V \quad = P$$
$$M_{max} \text{ (at fixed end)} = Pl$$
$$M_x = Px$$
$$\Delta_{max} \text{ (at free end)} = \frac{Pl^3}{3EI}$$
$$\Delta_x = \frac{P}{6EI}(2l^3 - 3l^2x + x^3)$$

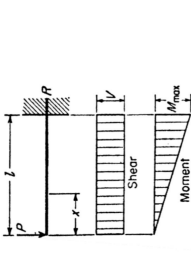

(p)

FIGURE 3.3 (*Continued*) Elastic-curve equations for prismatic beams. (*p*) Cantilever beam—concentrated load at free end.

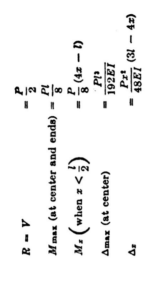

$$R = V \qquad\qquad = \frac{P}{2}$$

$$M_{max} \text{ (at center and ends)} = \frac{Pl}{8}$$

$$M_x \left(\text{when } x < \frac{l}{2}\right) = \frac{P}{8}(4x - l)$$

$$\Delta_{max} \text{ (at center)} \qquad = \frac{Pl^3}{192EI}$$

$$\Delta_x \qquad\qquad = \frac{Px^2}{48EI}(3l - 4x)$$

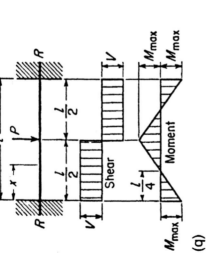

FIGURE 3.3 (*Continued*) Elastic-curve equations for prismatic beams. (*q*) Beam fixed at both ends—concentrated load at center.

SECTION 4
SHAFT AND SHAFTING FORMULAS

SHAFTS AND SHAFTING

When a shaft is subjected to a torque or twisting, a shearing stress is produced in the shaft. This shearing stress varies from zero at the shaft axis to a maximum at the outside surface or the extreme fiber. The relation between torque T, in pound-inches (kilonewton-meters), and the maximum shear stress s psi (MPa) is

$$T = s\,\frac{J}{c} \qquad (4.1)$$

where J is the polar moment of inertia of the cross section of the shaft, in⁴ (mm⁴), and c is the distance from the neutral axis to the extreme fiber.
For a solid circular shaft, c is equal to one-half the diameter and

$$J = \frac{\pi D^4}{32} \qquad (4.2)$$

where D = diameter. Thus,

$$s_s = \frac{TD/2}{\pi D^4/32} = \frac{5.1T}{D^3} \qquad (4.3)$$

For a hollow circular shaft

$$J = \frac{\pi(D^4 - d^4)}{32} \qquad (4.4)$$

where d = inside diameter. Thus

$$s_s = \frac{5.1TD}{D^4 - d^4} \qquad (4.5)$$

When you are selecting the diameter of a solid shaft for a known torque, use

$$D = 1.72 \sqrt[3]{\frac{T}{s}} \qquad (4.6)$$

where s = design stress of the shaft material. When the horsepower to be delivered and the shaft rotation speed are known, the diameter of the shaft can be determined from

$$D = 68.5 \sqrt[3]{\frac{H}{ns}} \qquad (4.7)$$

For hollow shafts in torsion only,

$$\frac{T}{s} = \frac{D^3(1 - q^4)}{5.10} \qquad (4.8)$$

where q = ratio of the inside diameter to the outside diameter. Also,

$$D = 1.72 \sqrt[3]{\frac{T}{s(1 - q^4)}} \qquad (4.9)$$

TORSIONAL DEFLECTION

When a shaft is transmitting torque from one end to the other, there is a tendency for the shaft to twist (Fig. 4.1). The total angle of twist in degrees for a solid circular shaft of uniform cross section is

$$\theta = \frac{584LT}{GD^4} \qquad (4.10)$$

For a tubular shaft

$$\theta = \frac{584LT}{G(D^4 - d^4)} \qquad (4.11)$$

where L = length of shaft, in (mm) and G = modulus of rigidity, psi (MPa).

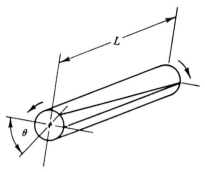

FIGURE 4.1 Torsional deflection of a machine shaft.

SHAFTS IN BENDING

When a shaft carriers only a bending load, it is treated as a beam. For a solid circular shaft subjected to a maximum bending moment M in pound-inches (kilonewton-meters), the maximum stress, psi (MPa), is

$$s_t = \frac{32M}{\pi D^3} \tag{4.12}$$

For a tubular shaft

$$s_t = \frac{32M}{\pi D^3}\frac{1}{1 - C^4} \tag{4.13}$$

where $C = d/D$.

SHAFTS WITH COMBINED TORSION AND BENDING

Shafts involved in the transmission of power by means of belts, gears, and chains are subjected not only to torsion but also to bending. To compute the effect of the combination of loads, the maximum shear stress theory is used for ductile metals, and the maximum normal stress theory is used for brittle metals. For maximum normal stress,

$$s_{t,\max} = \frac{s_t}{2} + \sqrt{s_s^2 + \frac{s_t^2}{4}} \tag{4.14}$$

where $s_{t,max}$ = maximum normal stress, psi (MPa)
$\qquad s_t$ = tensile stress, psi (MPa)
$\qquad s_s$ = shear stress, psi (MPa)

For maximum shear stress,

$$s_{s,max} = \sqrt{s_s^2 + \frac{s_t^2}{4}} \qquad (4.15)$$

where $s_{s,max}$ = maximum shear stress, psi (MPa). Thus, for a solid circular shaft,

$$s_{s,max} = \frac{5.1}{D^3} \sqrt{T^2 + M^2} \qquad (4.16)$$

where T = torque, lb·in (kN·m) and M = bending moment, lb·in (kN·m). And for a hollow circular shaft

$$s_{s,max} = \frac{5.1}{D^3} \sqrt{T^2 + M^2} \frac{1}{1 - C^4} \qquad (4.17)$$

According to the maximum normal stress theory, the maximum tensile stress due to the combined load is

$$s_{t,max} = \frac{s_t}{2} + \sqrt{s_s^2 + \frac{s_t^2}{4}} \qquad (4.18)$$

which for a solid circular shaft becomes

$$s_{t,max} = \frac{5.1}{D^3} (M + \sqrt{T^2 + M^2}) \qquad (4.19)$$

and for a hollow circular shaft

$$s_{t,max} = \frac{5.1}{D^3} (M + \sqrt{T^2 + M^2}) \left(\frac{1}{1 - C^4}\right) \qquad (4.20)$$

The term $\sqrt{T^2 + M^2}$ is often referred to as the *equivalent twisting moment,* and it is defined as the fictitious torsional moment that will induce the same shear stress in the shaft as the actual torsion and actual bending combined. For shafts with suddenly applied loads, a factor of 1.5 to 2.0 is applied to T and M, depending on the magnitude of the suddenly applied load. The higher values are used for larger loads.

POWER TRANSMISSION SHAFTING

Shafts made of ductile materials, based on strength, are designed using the maximum shear theory. The formulas below are based on ductile material shafts of circular cross section. Power transmission shafting is usually subjected to torsion, bending, and axial loads. For torsional loads, the torsional stress τ_{xy} is

$$\tau_{xy} = \begin{cases} \dfrac{M_t r}{J} = \dfrac{16 M_t}{\pi d^3} & \text{for solid shafts} & (4.21) \\[3ex] \dfrac{16 M_t d_o}{\pi (d_o^4 - d_i^4)} & \text{for hollow shafts} & (4.22) \end{cases}$$

For bending loads, the bending stress s_b (tension or compression) is

$$s_b = \begin{cases} \dfrac{M_b r}{l} = \dfrac{32 M_b}{\pi d^3} & \text{for solid shafts} & (4.23) \\[3ex] \dfrac{32 M_b d_o}{\pi (d_o^4 - d_i^4)} & \text{for hollow shafts} & (4.24) \end{cases}$$

For axial loads, the tensile or compressive stress s_a is

$$s_a = \begin{cases} \dfrac{4 F_a}{\pi d^2} & \text{for solid shafts} & (4.25) \\[3ex] \dfrac{4 F_a}{\pi (d_o^2 - d_i^2)} & \text{for hollow shafts} & (4.26) \end{cases}$$

The ASME Code equation for a hollow shaft combines torsion, bending, and axial loads by applying the maximum shear equation modified by introducing shock, fatigue, and column factors as follows:

$$d_o^3 = \frac{16}{\pi s_s (1 - K^4)} \sqrt{\left[K_b M_b + \frac{\alpha F_a d_o (1 + K^2)}{8} \right]^2 + (K_t M_t)^2} \quad (4.27)$$

For a solid shaft having little or no axial loading, the ASME Code equation reduces to

$$d^3 = \frac{16}{\pi s_s} \sqrt{(K_b M_b)^2 + (K_t M_t)^2} \quad (4.28)$$

where, at the section under consideration,

τ_{xy} = torsional shear stress, psi (MPa)
M_t = torsional moment, in·lb (kN·m)
M_b = bending moment, in·lb (kN·m)
d_o = shaft outside diameter, in (mm)
d_i = shaft inside diameter, in (mm)
F_o = axial load, lb (mm)
$K = \dfrac{d_i}{d_o}$
K_b = combineed shock and fatigue factor applied to bending moment
K_t = combined shock and fatigue factor applied to torsional moment

Combined Shock and Fatigue Factors Applied to Bending and Torsional Moments

	K_b	K_t
For stationary shafts:		
Load gradually applied	1.0	1.0
Load suddenly applied	1.5–2.0	1.5–2.0
For rotating shafts:		
Load gradually applied	1.5	1.0
Load suddenly applied (minor shock)	1.5–2.0	1.0–1.5
Load suddenly applied (heavy shock)	2.0–3.0	1.5–3.0

s_b = bending stress (tension or compression), psi (MPa)

s_a = axial stress (tension or compression), psi (MPa)

ASME Code specifies that for commercial steel shafting

s_s(allowable) = 8000 psi   for shaft without keyway   (55.2 MPa)

s_s(allowable) = 6000 psi   for shaft with keyway   (41.4 MPa)

ASME Code states that for steel purchased under definite specifications,

s_s(allowable) = 30 percent of elastic limit but not over 18 percent of ultimate strength in tension for shafts without keyways. These
values are to be reduced by 25 percent if keyways are present.

α = column action factor. The column action factor is unity for a tensile
load. For a compression load, α may be computed by

$$\alpha = \begin{cases} \dfrac{1}{1 - 0.0044(L/k)} & \text{for} \quad \dfrac{L}{k} < 115 \\[4mm] \dfrac{s_y}{\pi^2 nE}\left(\dfrac{L}{k}\right)^2 & \text{for} \quad \dfrac{L}{k} > 115 \end{cases}$$

$$n = \begin{cases} 1 & \text{for hinged ends} \\ 2.25 & \text{for fixed ends} \\ 1.6 & \text{for ends partly restrained as in bearings} \end{cases}$$

k = radius of gyration = $\sqrt{\dfrac{I}{A}}$ in (mm)

I = rectangular moment of inertia, in⁴ (mm⁴)

A = cross-sectional area of shaft, in² (mm²)

s_y = yield stress in compression, psi (MPa)

TORSIONAL RIGIDITY

Design of shafts for torsional rigidity is based on the permissible angle of twist. The amount of twist permissible depends on the particular application and varies about 0.08° per foot (0.3 m) for machine tool shafts to about 1.0° per foot (0.3 m) for line shafting.

$$\theta = \begin{cases} \dfrac{584 M_t L}{G(d^4_o - d^4_i)} & \text{for hollow circular shaft} & (4.29) \\[4mm] \dfrac{584 M_t L}{Gd^4} & \text{for solid circular shaft} & (4.30) \end{cases}$$

where θ = angle of twist, deg
    L = length of shaft, in (mm)
    M_t = torsional moment, in · lb (kN · m)
    G = torsional modulus of elasticity (MPa)
    d = shaft diameter, in (mm)

LATERAL RIGIDITY

Design of shafts for lateral rigidity is based on the permissible lateral deflection for proper bearing operation, accurate machine tool performance, satisfactory gear tooth action, shaft alignment, and other similar requirements. The amount of deflection may be determined by two successive integrations of

$$\frac{d^2y}{dx^2} = \frac{M_b}{EI} \tag{4.31}$$

where M_b = bending moment, in · lb (kN · m); E = modulus of elasticity psi (MPa); and I = rectangular moment of inertia, in⁴ (mm⁴). If the shaft is of variable cross section, a graphical solution of the above expression is practical.

MOMENTS

Bending and torsional moments are the main factors influencing shaft design. One of the first steps in shaft design is to draw the bending moment diagram for the loaded shaft or the combined bending moment diagram if the loads acting on the shaft are in more than one axial plane. From the bending moment diagram, the points of critical bending stress can be determined.

The torsional moment acting on the shaft can be determined from

$$M_t = \frac{\text{hp} \times 33,000 \times 12}{2\pi \text{ rpm}} = \frac{63,000 \times \text{hp}}{\text{rpm}} \qquad \text{in · lb (kN · m)} \tag{4.32}$$

For a belt drive, the torque is found from

$$M_t = (T_1 - T_2)R \qquad \text{in · lb (kN · m)} \tag{4.33}$$

where T_1 = tight side of belt on pulley, lb; T_2 = loose side of belt on pulley, lb (kg); and R = radius of pulley, in (mm).

For a gear drive, the torque is found from

$$M_t = F_t R \tag{4.34}$$

where F_t = tangential force at the pitch radius, lb (kg), and R = pitch radius, in (mm).

CRITICAL SPEED

If a shaft has a concentrated load located somewhere along its span, the critical speed is

$$N_c = \frac{188}{\sqrt{Y}} \tag{4.35}$$

where N_c = critical speed, rpm, and Y = deflection of the shaft, in (mm). The deflection is found by using the methods for beams.

If the shaft is steel, has a solid circular cross section, and is supported by thin bearings or self-aligning bearings, then

$$N_c = 387,000 \frac{D^2}{ab} \sqrt{\frac{L}{P}} \qquad (4.36)$$

where D = diameter, in (mm); L = distance between supporting bearings, in (mm); P = load, lb (kg); and a and b = distance from load to bearings, in (mm). If the shaft is rigidly supported in long bearings,

$$N_c = 387,000 \frac{D^2 L}{ab} \sqrt{\frac{L}{Pab}} \qquad (4.37)$$

For these equations, the shaft must be of almost uniform diameter—small shoulders and reliefs may be ignored. If the weight of the shaft is relatively small, it is usually ignored; but if it is to be included, one-half the weight is added to the load.

For most applications, the maximum shaft operating speed should not exceed 80 percent of the critical speed.

For a shaft of constant cross section, simply supported at the ends, with no mass involved other than that of the shaft itself, the first critical speed is very nearly

$$\omega_c = \sqrt{\frac{5}{4} \frac{g}{\delta(\max)}} \qquad \text{rad/unit time} \qquad (4.38)$$

where $\delta(\max)$ is the maximum static deflection caused by a uniformly distributed load equal to the weight of the shaft and g is the gravitational constant (32.2 ft/s^2 or 386 in/s^2).

For a shaft of negligible mass carrying several concentrated masses (see Fig. 4.2) the first critical speed is approximately

$$\omega_c = \sqrt{\frac{g \sum_1^j W_n \delta_n}{\sum_1^j W_n \delta_n^2}} \qquad \text{Rayleigh-Ritz equation} \qquad (4.39)$$

where W_n = weight of nth mass; δ_n = static deflection at nth mass; and j = total number of masses.

This same equation can be used for estimating the first critical speed of a shaft with distributed mass. Refer to Fig. 4.3. Break the distributed mass

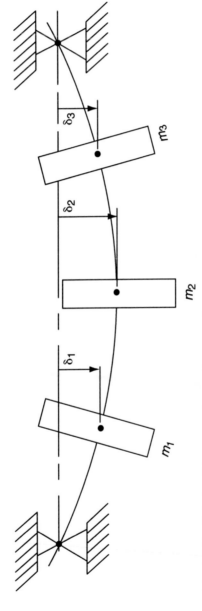

FIGURE 4.2 Shaft of neglible mass carrying several concentrated masses.

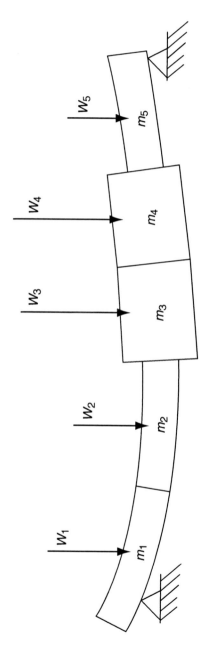

FIGURE 4.3 Shaft with a distributed mass.

into a number of pieces m_1, m_2, m_3, etc. Treat the mass of each piece as though it were concentrated at its center of gravity.

The Dunkerley equation, another approximation for the first critical speed of a multimass system, is

$$\frac{1}{\omega_c^2} = \frac{1}{\omega_1^2} + \frac{1}{\omega_2^2} + \frac{1}{\omega_3^2} + \cdots \qquad \text{Dunkerley equation} \qquad (4.40)$$

where ω_c is the first critical speed of the multimass system; ω_1 is the critical speed which would exist if only mass 1 were present; ω_2 is the critical speed with only mass 2, etc.

It is important to keep in mind that both the Rayleigh-Ritz and the Dunkerley equations are approximations to the first natural frequency of vibration, which is assumed to be nearly equal to the critical speed of rotation. In general, the Rayleigh-Ritz equation overestimates and the Dunkerley equation underestimates the natural frequency.

EMPIRICAL FORMULAS FOR STEEL TRANSMISSION SHAFTING

A number of empirical formulas for power transmission shafting are regularly used. Most such formulas include one or more service factors.

For solid shafts,

D = outside diameter of shaft, in (mm)
T = maximum torsion moment, in·lb (N·m)
B = maximum bending moment in inch pounds (Nm)
K_t = service factor applied to T; ranges between 1.0 and 2.0†
K_b = service factor applied to B; ranges between 1.0 and 2.5†
π = constant, 3.1416
hp = horsepower transmitted (kW)
R = revolutions per minute
S = allowable working stress, psi (MPa)

For combined torsion and bending,

$$D = \sqrt[3]{\frac{16}{\pi Sc} \sqrt{(K_t T)^2 + (K_b B)^2}} \qquad (4.41)$$

or

---

† See an engineering handbook for the exact value.

$$\sqrt[3]{\frac{16}{\pi Sc}} \sqrt{\left(\frac{396,000K_t \text{ hp}}{2\pi R}\right)^2 + (K_b B)^2} \qquad (4.42)$$

For bending only,

$$D = \sqrt[3]{\frac{32K_b B}{\pi S_b}} \qquad K_b B = \frac{\pi}{32} S_b D^3 = 0.09817 S_b D^3 \quad (4.43)$$

For torsion only,

$$D = \sqrt[3]{\frac{321,000K_t \text{ hp}}{S_t R}} \qquad K_t T = 0.1963 S_t D^3$$

$$(4.44)$$

$$K_t \text{ (hp)} = \frac{S_t D^3 R}{321,000}$$

For equivalent hollow shafts, find D for the solid shaft and multiply by $1/\sqrt[3]{1 - K^4}$ to give the outside diameter of the equivalent hollow shaft of any desired ratio K, where D_1 = inside diameter of hollow shaft and K = ratio of inside diameter to outside diameter = D_1/D.

POWER TRANSMISSION "GROUP-SYSTEM" SHAFTING

Figure 4.4 gives the names and locations of the various shafts used in the group system of driving belting, chains, or a combination of these media.

Head Shafts

Location: First shaft from motor or prime mover.
Speeds: Industrial range 25 to 550 rpm.

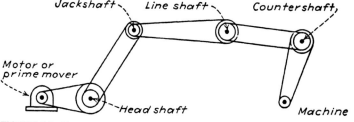

FIGURE 4.4 Group-system shafting.

Formula:

$$\text{hp} = \frac{D^3 R}{125} \qquad (4.45)$$

$$D = \sqrt[3]{\frac{125 \times \text{hp}}{R}} \qquad (4.46)$$

where hp = horsepower (kW)
  D = shaft diameter, in (mm)
  R = rotation speed, rpm
  125 = constant for combined torsional and bending safe-working stress of 2800 psi (19.3 MPa)

Line Shafts

Location: It can be first shaft from motor or first or second shaft from head shaft.

Speeds: Industrial range 70 to 400 rpm.

Formula:

$$\text{hp} = \frac{D^3 R}{100}$$

$$D = \sqrt[3]{\frac{100 \times \text{hp}}{R}} \qquad (4.47)$$

where hp = horsepower (kW)
  D = shaft diameter, in (mm)
  R = rotation speed, rpm
  100 = constant for a combined torsional and bending safe-working stress of 3200 psi (22.1 MPa)

Jackshafts

Location: Either between head and line shafts, or between line and countershafts.

Speeds: Industrial range 100 to 400 rpm.

Formula: Same as for line shafting, Eq. (4.47)

Countershafts

Location: Usually between line shaft and driven apparatus or machine.
Speeds: Industrial range 100 to 600 rpm.
Formula: For average service and loading

$$\text{hp} = \frac{D^3 R}{80} \qquad (4.48)$$

$$D = \sqrt[3]{\frac{80 \times \text{hp}}{R}} \qquad (4.49)$$

where hp = horsepower (kW)
  D = shaft diameter, in (mm)
  R = rotation speed, rpm
  80 = constant for combined torsional and bending safe-working stress of 3600 psi (24.8 MPa)

Table 4.1 gives the angle-of-twist θ formulas for shafts of various cross sections, while Table 4.2 gives shaft torque formulas and the location of the maximum shear stress in the shaft.

FLYWHEELS ON SHAFTS

Some power sources, such as internal combustion engines, produce energy during a small portion of the cycle. A flywheel is used to smooth out these fluctuations and to make the flow of energy uniform.
  The kinetic energy (KE) in a flywheel is

$$\text{KE} = \frac{W v^2}{2g} \qquad (4.50)$$

where KE = kinetic energy, ft · lb (kN · m)
  W = weight of flywheel, lb (kg)
  v = velocity of center of mass ft/s (m/s)
  g = acceleration due to gravity, ft/s^2 (m/s^2)

The acceptable variation in angular velocity depends on the application. The coefficient of regulation is used to specify the magnitude of the variation in angular velocity. It is found with

$$C_f = \frac{v_1 - v_2}{v} \qquad (4.51)$$

where C_f = coefficient of regulation.

TABLE 4.1 Shaft Angle-of-Twist Section Formulas

θ = twist, radians
T = torque, in-lb (N-m)
L = length of shaft, in (mm)
N = modulus of rigidity, psi (MPa)
D, d_o, d_i, d_m, d_s, s, a, b = shaft sectional dimensions, in (mm)

Shaft section	Angle-of-twist θ =
D	$\dfrac{32TL}{\pi D^4 N}$
d_i d_o	$\dfrac{32TL}{\pi(d_0^4 - d_1^4)N}$
d_s d_m	$\dfrac{16(d_m^2 + d_s^2)TL}{\pi d_m^3 d_s^3 N}$
s	$\dfrac{7.11TL}{s^4 N}$
a b	$\dfrac{3.33(a^2 + b^2)TL}{a^3 b^3 N}$

$$\Delta KE = \frac{WC_f v^2}{g} \qquad (4.52)$$

If the rim thickness is small relative to the diameter (as is usually the case), the center of the rim can be taken as the center of mass. The acceleration due to gravity is usually taken as 32.2.

TABLE 4.2 Shaft Torque Formulas and Location of Maximum Shear Stress in the Shaft

Shaft section	Location of max shear	Torque formulas: $T =$
D	Outer fiber	$\dfrac{\pi D^3 f}{16}$
d_i d_o	Outer fiber	$\dfrac{\dfrac{\pi}{16}(d_o^4 - d_i^4)}{d_o} f$
d_s d_m	Ends of minor axis	$\dfrac{\pi d_m d_s^2 f}{16}$
S	Middle of sides	$0.208 S^3 f$
A B	Midpoint of major sides	$\dfrac{A^2 B^2 f}{3A + 1.8B}$

Note: f = maximum shear stress, psi (MPa)

$$W = \frac{32.2 \ \Delta KE}{C_f v^2} \qquad (4.53)$$

where W = weight of rim, lb (kg), and v = average velocity of the center of the rim, ft/s (m/s).

VERTICAL SHAFTS FOR MIXING OR AGITATION VESSELS

Horsepower formula

$$hp \ required = \frac{WV^2}{550G} \qquad (4.54)$$

where W = total weight of material being mixed or agitated
      V = velocity, ft/s (figure mean diameter of paddles)
      G = gravity, or 32.2

Note: Horsepower (hp) $\times$ 0.746 = kW.

MACHINE COMPONENT AND RELIABILITY FORMULAS

SPRINGS

Helical Springs

For helical compression or tension springs (Fig. 5.1),

$$\tau = \frac{8PD}{\pi d^3} \tag{5.1}$$

$$P = \frac{\pi d^3 \tau}{8D} \tag{5.2}$$

$$\delta = \frac{8PD^3 n}{Gd^4} \tag{5.3}$$

$$\delta = \frac{\pi D^2 n \tau}{Gd} \tag{5.4}$$

$$P = \frac{Gd^4 \delta}{8D^3 n} \tag{5.5}$$

$$k = \frac{P}{\delta} = \frac{Gd^4}{8D^3 n} \tag{5.5a}$$

$$\tau = \frac{\delta G d}{\pi D^2 n} \tag{5.6}$$

$$\tau' = K\tau = \frac{8PDK}{\pi d^3} \tag{5.7}$$

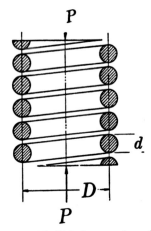

FIGURE 5.1   Helical compression spring.

$$K = \frac{4c - 1}{4c - 4} + \frac{0.615}{c} \tag{5.8}$$

$$c = \frac{D}{d} = \text{spring index}$$

where P = load on spring (Fig. 5.1)
   d, D = bar and coil diameters
      δ = deflection
      τ = uncorrected shear stress
     $\tau' = K\tau$ = corrected shear stress
      K = curvature correction factor (given in engineering handbooks)
      n = number of active coils in the spring
      $k = P/\delta$ = spring rate or stiffness

For a compression spring with ends prevented from unwinding during deflection, the expansion in diameter ΔD during compression from free to solid height is

$$\Delta D = 0.05 \frac{p^2 - d^2}{D} \tag{5.9}$$

where p = pitch, or center-to-center distance, of coils at free height. If the ends are free to unwind, the expansion in diameter ΔD becomes

$$\Delta D = 0.10 \frac{p^2 - 0.8pd - 0.2d^2}{D} \tag{5.10}$$

Lateral Loading of Compression Springs

Frequently helical springs, e.g., when used as vibration isolators, are laterally loaded by a force F while being compressed by a vertical force P, the only resistance to lateral deflection being the stiffness of the spring (Fig. 5.1).

For steel springs of round wire with $E = 30 \times 10^6$ psi (206.9 GPa) and $G = 11.5 \times 10^6$ psi (79.3 GPa), the stiffness k_x in the lateral direction is given by

$$k_x = \frac{F}{\delta_x} = \frac{10^6 d^4}{C_l n D (0.204 h_s^2 + 0.265 D^2)} \tag{5.11}$$

where δ_x = lateral deflection due to force F (Fig. 5.2)
l_0 = free length
h_s = compressed length of spring = $l_0 - \delta_{st}$
δ_{st} = vertical deflection due to load P
C_t = a factor depending on δ_{st}/l_0 and l_0/D

Values of C_l may be taken from charts in engineering handbooks.

The ratio of axial stiffness $k_y = P/\delta_{st}$ to lateral stiffness k_x for steel springs of round wire with $E = 30 \times 10^6$ psi, (206.9 GPa) and $G = 11.5 \times 10^6$ psi (79.3 GPa) is obtained by using

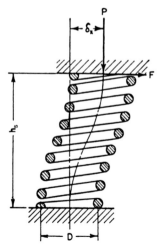

FIGURE 5.2 Spring under combined axial and lateral loading (one end fixed, the other guided).

$$\frac{k_y}{k_x} = 1.44C_l \left(0.204 \frac{h_s^2}{D^2} + 0.265\right) \tag{5.12}$$

This equation will also apply to other spring materials where the ratio E/G is approximately 2.6.

Natural Frequencies. The lowest natural frequency f_n in cycles per second or hertz (Hz), for a helical spring clamped beween two parallel plates is

$$f_n = \frac{2d}{\pi D^2 n} \sqrt{\frac{Gg}{32\gamma}} \tag{5.13}$$

where n = number of active coils and g = acceleration of gravity.

For steel springs having both ends clamped where $G = 11.5 \times 10^6$ psi (79.3 GPa) and $\gamma = 0.285$ lb/in³ (0.008 kg/cm³), the frequency f_n (Hz) becomes

$$f_n = \frac{14,000d}{D^2 n} \tag{5.14}$$

Frequencies in the higher modes of vibration are 2, 3, 4, etc., times this frequency.

Impact Effects. If one end of a long precompressed or free helical spring is suddenly compressed by a heavy mass moving with a velocity v, a surge wave is propagated along the spring *wire* with a velocity v_s, where

$$v_s = \frac{d}{D} \sqrt{\frac{gG}{2\gamma}} \tag{5.15}$$

For steel with $G = 11.5 \times 10^6$ psi (79.3 GPa) and $g = 386$ in/s (980.4 cm/s),

$$\gamma = 0.283 \text{ lb/in}^3 \text{ (7833 kg/m}^3\text{)}$$

$$v_s = \frac{88,560d}{D} \qquad \text{in/s (cm/s)} \tag{5.16}$$

The surge time in seconds for the wave to cover the whole length of the spring wire is

$$t_s = \frac{\pi n D}{v_s} \qquad \text{s} \tag{5.17}$$

The increment in uncorrected shear stress $\Delta\tau$, when one end of a long spring is suddenly compressed with a velocity v, is

$$\Delta\tau = v\sqrt{\frac{2\gamma G}{g}} \qquad (5.18)$$

For steel springs, with $G = 11.5 \times 10^6$ psi (79.3 GPa) and $\gamma = 0.285$ lb/in^3 (7888 kg/m^3),

$$\Delta\tau = 130v \qquad (5.19)$$

where v is given in inches per second (centimeters per second) and $\Delta\tau$ in pounds per square inch (megapascals). This indicates that $\Delta\tau$ is independent of the spring dimensions.

The corresponding increments in load ΔP and deflection per coil $\Delta\delta$ are

$$\Delta P = \frac{\pi v d^3}{8D}\sqrt{\frac{2\gamma G}{g}} \qquad (5.20)$$

$$\Delta\delta = \frac{\pi v D^2}{d}\sqrt{\frac{2\gamma}{gG}} \qquad (5.21)$$

For steel springs, with ΔP in pounds (newtons), $\Delta\delta$ in inches (centimeters), and v in inches per second (centimeters per second),

$$\Delta P = \frac{51 d^3 v}{D} \qquad (5.22)$$

$$\Delta\delta = \frac{D^2 v}{d}(35.5 \times 10^{-6}) \qquad (5.23)$$

Helical Extension Springs

Approximate formulas for tension springs with their half loop turned up to form the end coil (Fig. 5.3) are as follows:

$$\text{Bending moment at } A' = \frac{PD}{2}$$

where D = the mean coil diameter.

$$\text{Nominal bending stress at } A' = \frac{16PD}{\pi d^3}$$

where d is the wire diameter.

$$\text{Nominal tension stress due to torsion moment} = \frac{8PD}{\pi d^3}$$

The maximum bending stress σ at point A' (at the start of the bend) is

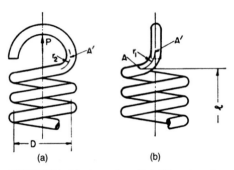

FIGURE 5.3   Tension spring with half loop turned up to form end coil.

$$\sigma = \frac{16PD}{\pi d^3} K_1 + \frac{4P}{\pi d^2} \tag{5.24}$$

A value of $K_1 = r_o/r_i$ (where r_o and r_i are the mean and inside radii, respectively, of the sharp bend) is suggested by the ASM committee for calculating bending stress.

The maximum, stress τ_1 due to the torsion moment Pr will then be (approximately)

$$\tau_1 = \frac{8PD}{\pi d^3} \frac{4c_1 - 1}{4c_1 - 4} \tag{5.25}$$

The amount of initial tension which can be put into an extension spring depends primarily on the spring index D/d; and in general, the higher the index, the lower the initial tension loads. Values of initial tension load P_1 which can be obtained in practice are given by

$$P_1 = \frac{\pi \tau d^3}{8D} \tag{5.26}$$

where τ = uncorrected torsion stress resulting from initial tension. Values of τ which may be obtained in practice are given in engineering handbooks.

Square and Rectangular-Bar Helical Springs

Typical rectangular wire and square wire springs are shown in Figs. 5.4 to 5.6. An estimate of the amount of upsetting due to coiling of a rectangular or square wire section can be obtained from the formula

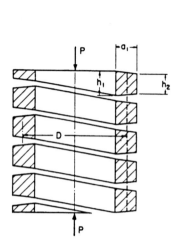

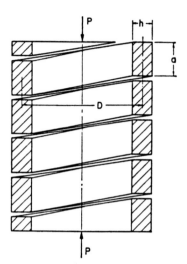

FIGURE 5.4 Helical spring of square wire axially loaded. (*Note:* The wire becomes somewhat trapezoidal during the coiling operation.)

FIGURE 5.5 Helical spring with long side of bar parallel to axis.

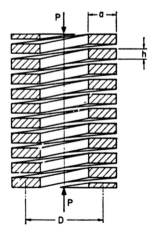

FIGURE 5.6 Helical spring of rectangular wire, coiled flatwise.

$$h_1 = h \left[1 + \frac{k(D_o - D_i)}{D_o + D_i} \right] \tag{5.27}$$

where D_o, D_i = outside and inside diameters, respectively, of spring
$\quad\quad\quad h$ = original thickness
$\quad\quad\quad h_1$ = upset thickness (Fig. 5.4)
$\quad\quad\quad k = \begin{cases} 0.3 & \text{for cold-wound springs} \\ 0.4 & \text{for hot-wound springs and annealed materials} \end{cases}$

Uncorrected Stress. The uncorrected stress τ in a square wire spring is obtained by assuming that the spring acts essentially as a straight bar under torsion. This gives

$$\tau = \frac{2.4PD}{a^3} \tag{5.28}$$

where P = load
$\quad\quad\quad D$ = mean coil diameter
$\quad\quad\quad a$ = side of square cross section

Where the section becomes trapezoidal (Fig. 5.4) the quantity a in Eq. (5.28) may be taken as the average length of a side $(2a_1 + h_1 + h_2)/4$.

Corrected Stress. The corrected stress τ', which includes effects of curvature and direct shear and which should be used to calculate the stress range for fatigue loading, is given by

$$\tau' = K'\tau \tag{5.29}$$

where $\quad\quad K' = 1 + \frac{1.2}{c} + \frac{0.56}{c^2} + \frac{0.5}{c^3} \tag{5.30}$

In this case the spring index $c = D/a$ or D/a_1. Values of the curvature correction factor K' are given in engineering handbooks. This factor is slightly below the factor K for round wire and applies for $c = D/a$ greater than 3.

Deflection. The deflection δ for a square bar helical spring is given by

$$\delta = \frac{5.59PD^3n}{Ga^4} \tag{5.31}$$

where n = number of active coils and G = modulus of rigidity. This formula is theoretically around 2 to 4 percent in error for springs with indexes between 3 and 4, but for most practical cases it is accurate enough. Somewhat more accurate results may be obtained by using

$$\text{Spring rate} = \frac{P}{\delta} = \frac{Ga^4}{5.59D^3n} \tag{5.32}$$

Uncorrected Stress.   For rectangular bar springs (Figs. 5.4 and 5.5) the uncorrected stress τ is given by

$$\tau = \frac{PD}{k_1ah^2} \tag{5.33}$$

where a = long side of bar cross section
$\quad h$ = short side of bar cross section
$\quad k_1$ = a factor depending on a/h which is given in engineering handbooks

Stress τ', corrected, is

$$\tau' = \beta \frac{PD}{ab\sqrt{ab}} \tag{5.34}$$

where β is a factor taken from engineering handbooks.

Deflection (Large-Index Springs).   For springs of large index D/a or D/h, the deflection δ may be calculated by assuming the spring to act essentially as a straight bar, with sides a and h, subjected to a torsion moment $PD/2$. This gives

$$\delta = \frac{PD^3n}{k_2ah^3G} \tag{5.35}$$

where n = number of active coils
$\quad G$ = modulus of rigidity
$\quad k_2$ = a factor depending on a/h which may be taken from handbooks.

Note that $h < a$.

The spring rate constant or rate k in pounds per inch (kilograms per centimeter) is given by

$$k = \frac{P}{\delta} = \frac{a^2b^2G}{D^3n\gamma} \tag{5.36}$$

Helical Torsion Springs

Round-Wire Springs.   In design it is common practice to consider the spring as acted on by a moment M acting about the helix axis. The following formulas apply for round-wire springs (Fig. 5.7).

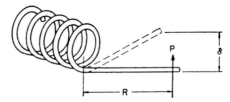

FIGURE 5.7 Typical method of loading torsion spring.

$$\sigma = \frac{10.2M}{d^3} \qquad (5.37)$$

$$\sigma' = K_1\sigma = K_1 \frac{10.2M}{d^3} \qquad (5.38)$$

Here K_1 is taken as a function of D/d. Angular deflection φ turns, due to moment M, are

$$\varphi = \frac{10.2MnD}{Ed^4} \qquad \text{turns} \qquad (5.39)$$

Angular deflection φ_1 degrees, due to M, is

$$\varphi_1 = \frac{3670MnD}{Ed^4} \qquad \text{deg} \qquad (5.40)$$

Spring rate or constant k, in inch-pounds per degree (centimeter-newtons per degree), is

$$k = \frac{M}{\phi_1} = \frac{Ed^4}{3670nD} \qquad (5.41)$$

Active length of wire l required is

$$l = \frac{Ed^4}{1170k} \qquad (5.42)$$

where σ, σ' = uncorrected and corrected stresses, respectively
  D, d = mean coil and wire diameters, respectively
  E = modulus of elasticity
  n = number of active coils

Square-Wire Springs

$$\sigma = \frac{6M}{h^3} \tag{5.43}$$

$$\sigma' = K_2\sigma = K_2\frac{6M}{h^3} \tag{5.44}$$

$$\varphi = \frac{6MnD}{Eh^4} \quad \text{turns} \tag{5.45}$$

$$\varphi_1 = \frac{2160MnD}{Eh^4} \quad \text{deg} \tag{5.46}$$

where h = side of square cross section
$\quad D$ = mean coil diameter
$\quad K_2$ = a function of D/h, taken from engineering handbooks

Rectangular Bar Springs

$$\sigma = \frac{6M}{bh^2} \tag{5.47}$$

$$\sigma' = K_2\sigma = K_2\frac{6M}{bh^2} \tag{5.48}$$

$$\varphi = \frac{6MnD}{Ebh^3} \quad \text{turns} \tag{5.49}$$

$$\varphi_1 = \frac{2160MnD}{Ebh^3} \quad \text{deg} \tag{5.50}$$

where h = radial depth of rectangular cross section
$\quad D$ = mean coil diameter
$\quad b$ = width of rectangular cross section (in axial direction)
$\quad K_2$ = a function of D/h, taken from engineering handbooks

If a load P acts on an arm at a distance R from the axis, the moment M in the above equations may be taken to equal PR. In this case the deflection in a circumferential direction at radius R will be $\phi_1 R/57.3$.

SPIRAL POWER, AND NEG'ATOR SPRINGS

Hairsprings generally have a large number of turns not in contact; one end is clamped. The number of revolutions or turns n produced by a moment M is (Fig. 5.8)

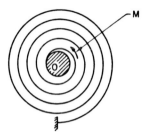

FIGURE 5.8 Spiral spring with large number of turns (clamped outer end).

$$n = \frac{6Ml}{\pi Ebh^3} \qquad (5.51)$$

where h = thickness of strip
  b = width
  l = active length of strip
  E = modulus of elasticity

The angular deflection in degrees in $360n$.
  The bending stress σ is given by

$$\sigma = \frac{6M}{bh^2} \qquad (5.52)$$

or

$$\sigma = \frac{\pi nhE}{l} \qquad (5.53)$$

Power or motor springs (Fig. 5.9), such as in a clock, may be placed inside a hollow case. If l is the active length of spring strip and h is the thickness, the total sectional area of the wound spring will be lh. Then,

$$d_2 = \sqrt{\frac{4}{\pi} lh + d_1^2} \qquad (5.54)$$

where d_2 = outer diameter of wound-up spring and d_1 = arbor diameter. Also assuming that adjacent coils touch, the number of turns n for the wound-up condition becomes

$$n = \frac{d_2 - d_1}{2h} \qquad (5.55)$$

The total number of turns ΔN delivered by the spring in unwinding from the wound position of Fig. 5.10 to the unwound position of Fig. 5.9 will be the difference between n and n'. Thus

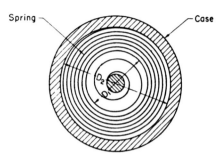

FIGURE 5.9 Power spring unwound and resting inside case.

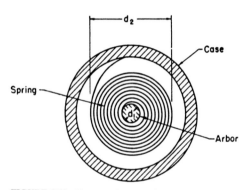

FIGURE 5.10 Power spring wound on arbor.

$$\Delta N = n - n'$$

$$= \frac{\sqrt{(4/\pi)lh + d_1^2} + \sqrt{D_2^2 - (4/\pi)lh} - (D_2 + d_1)}{2h} \quad (5.56)$$

$$l = \frac{D_2^2 - d_1^2}{2.55h} \quad (5.57)$$

and

$$\Delta N = \frac{D_2^2 - d_1^2}{2hU} = \frac{4l}{\pi U} \quad (5.58)$$

where

$$U = \frac{D_2^2 - d_1^2}{\sqrt{2(D_2^2 + d_1^2)} - (D_2 + d_1)} \quad (5.59)$$

The formulas for stress σ, moment M, and total number of turns ΔN, obtained by assuming the strip to be loaded by a constant moment along its length, are

$$M = \frac{\sigma b h^2}{6} \tag{5.60}$$

$$\sigma = \frac{6M}{bh^2} = \frac{\pi E h\, \Delta N}{l} \tag{5.61}$$

$$\Delta N = \frac{6Ml}{\pi E b h^3} = \frac{\sigma l}{\pi E h} \tag{5.62}$$

Also

$$h = \sqrt[3]{\frac{1.5MU}{Eb}} \tag{5.63}$$

In these b = width of strip, h = thickness, and E = modulus of elasticity.

Constant-Force Spring (Neg'ator)

For such a spring the load P is given by

$$P = \frac{Ebh^3}{26.4} \left[\frac{1}{R_y^2} - \left(\frac{1}{R_n} - \frac{1}{R_1} \right)^2 \right] \tag{5.64}$$

where b, h = width and thickness of strip, respectively
E = modulus of elasticity
R_n = minimum natural radius of curvature of coil
R_1 = outer radius of coil

Design Formulas—Neg'ator Extension Springs. In designing Neg'ator extension spring (Fig. 5.11) the following formulas, given by Votta, may be used.

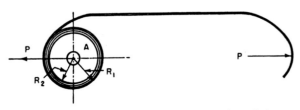

FIGURE 5.11 Constant-force spring (Neg'ator extension spring).

For Springs with 10 Coils or Fewer

$$h \geq \frac{26.4P}{EbS_f^2} \tag{5.65}$$

$$R_n = \sqrt{\frac{Ebh^3}{26.4P}} \tag{5.66}$$

$$b = \frac{26.4P}{EhS_f^2} \tag{5.67}$$

$$R_2 = 1.15R_n \tag{5.68}$$

$$L = \delta + 10R_2 \tag{5.69}$$

where P = load
R_n = minimum natural radius of curvature of coil
L = total length of Neg'ator
R_2 = radius of storage bushing
δ = deflection required
S_f = a factor depending on number of cycles of operation

For Springs with More than 10 Coils

$$h \geq \frac{26.4P}{EbS_f^2} \tag{5.70}$$

$$R_m = \sqrt{\frac{Ebh^3}{26.4P}} \tag{5.71}$$

$$R_n = \frac{R_m}{1.15} \tag{5.72}$$

$$R_2 = 1.15R_m \tag{5.73}$$

$$L = \delta + 10R_2 \tag{5.74}$$

where R_m = maximum natural radius or curvature of coil.

CONED-DISK OR BELLEVILLE SPRINGS

In many designs, the use of coned-disk springs (also known as *Belleville springs*) is an advantage. Such springs consist essentially of circular disks dished to a conical shape, as shown in the diametral cross section of

Fig. 5.12. When load is applied as indicated, the disk tends to flatten out, and this elastic deformation constitutes the spring action.

Load is assumed to act at the edges.

$$P = \frac{C_1 C E t^4}{R^2} \tag{5.75}$$

where P = load at deflection δ from no-load position
C = factor depending on R/r
R, r = outside and inside radii, respectively
C_1 = factor depending on δ/t and h/t from engineering handbooks
h = initial cone height of spring
t = thickness

Elastic stresses at deflection δ (minus signs indicate compressive stress) are

$$\sigma_c = -K_c \frac{E t^2}{R^2} \tag{5.76}$$

$$\sigma_{t1} = K_{t1} \frac{E t^2}{R^2} \tag{5.77}$$

$$\sigma_{t2} = K_{t2} \frac{E t^2}{R^2} \tag{5.78}$$

where E = modulus of elasticity
σ_c = stress at upper inner edge A
σ_{t1} = stress at lower inner edge B
σ_{t2} = stress at lower outer edge C

The factors K_c, K_{t1}, and K_{t2} depend on δ/t, h/t, and R/r, respectively. For any h/t and δ/t the factor K_c may be taken from the charts in engineering handbooks.

Nominal Stress σ_n.   This stress is also called *simple bending stress* and is used for static loading.

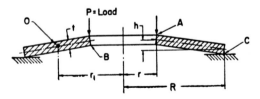

FIGURE 5.12   Coned-disk spring (Belleville spring).

$$\sigma_n = \frac{0.96P}{t^2} \tag{5.79}$$

Formulas for Small Cone Height and Deflection

These formulas are based on elastic flat-plate theory and assume δ/t and h/t are less than 0.5; also, load is assumed to act at the edges.

$$P = K_1 \frac{\delta E t^3}{R^2} \tag{5.80}$$

$$P = \frac{\sigma t^2}{K_3} \tag{5.81}$$

$$\sigma = K_2 \frac{\delta E t}{R^2} \tag{5.82}$$

$$\sigma = K_3 \frac{P}{t^2} \tag{5.83}$$

$$\delta = \frac{PR^2}{K_1 E t^3} \tag{5.84}$$

where σ = stress at inner edge. In these the factors K_1, K_2, and K_3 depend on R/r.

Load P′ Acting inside Edges

$$P' = P \frac{R - r}{a} \tag{5.85}$$

$$\delta' = \delta \frac{a}{R - r} \tag{5.86}$$

where a = radial distance between rims
$\quad P'$ = load on rims
$\quad \delta'$ = deflection between rims

The load P is calculated for a given deflection δ between inside and outside edges. Stresses δ_c, σ_{t1}, and σ_{t2} are calculated from δ/h and h/t.
Nominal stress σ_n is

$$\sigma_n = 0.96 \frac{P'}{t^2} \frac{a}{R - r}$$

FLAT AND LEAF SPRINGS

Simple Cantilever Spring—Constant-Width

Small Deflections. The simplest type of flat spring is the simple cantilever spring loaded by an end load, as in Fig. 5.13. In this case the well-known cantilever formula for deflection is

$$\delta = \frac{Pl^3}{3EI} \tag{5.87}$$

where l = length of spring
E = modulus of elasticity
I = moment of inertia of spring cross section ($I = b_0 h^3 / 12$ where b_0 = width, h = thickness)

For cases where the strip width is large compared to the thickness, that is, b_0/h large, and where b_0 is comparable to l, it is frequently more accurate to take I equal to $b_0 h^3 / 12(1 - \nu^2)$ (ν = Poisson's ratio).

$$\delta = \frac{Pl^3}{3EI} (1 - \nu^2) \tag{5.88}$$

For most metals, ν is around 0.3, which means that in such cases the spring will be about 10 percent stiffer than calculated. In most practical cases the deflection will probably lie between the results calculated.

The nominal bending stress σ at the built-in edge of the spring of Fig. 5.13 is

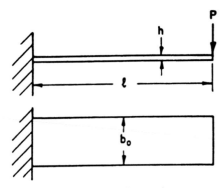

FIGURE 5.13 Simple cantilever spring.

$$\sigma = \frac{6Pl}{b_0 h^2} \qquad (5.89)$$

Simple Cantilever Springs—Trapezoidal Profile

Small Deflections. In many cases, leaf springs of the usual shape may be considered as cantilever springs of trapezoidal profile (Fig. 5.14). From beam theory, the deflection is

$$\delta = K_1 \frac{Pl^3}{3EI_0} \qquad (5.90)$$

where

$$K_1 = \frac{3}{(1 - b/b_0)^3} \left[\frac{1}{2} - 2\frac{b}{b_0} + \left(\frac{b}{b_0} \right)^2 \left(\frac{3}{2} - \ln \frac{b}{b_0} \right) \right] \qquad (5.91)$$

and I_0 = moment of inertia at the built-in end. Factor K_1 depends on b/b_0 and may be taken from engineering handbooks.

Flat Springs under Combined Axial and Lateral Loading

Frequently in practice flat springs are loaded as indicated in Fig. 5.15, with one end being built-in and the other being free to move laterally but restrained from rotation. If the axial load P is small compared with the buckling load, the deflection and stress are

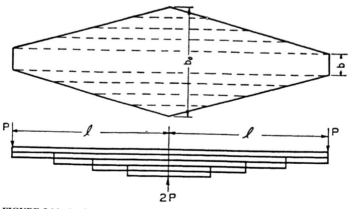

FIGURE 5.14 Leaf spring equivalent to trapezoidal cantilever spring.

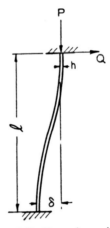

FIGURE 5.15 Flat spring under combined axial and transverse loading.

$$\delta = \frac{Ql^3}{12EI} \tag{5.92}$$

$$\sigma = \frac{3\delta Eh}{l^2} \tag{5.93}$$

where l = length of beam
$\quad Q$ = lateral load
$\quad I = bh^3/12$ = moment of inertia of section
$\quad b$ = width
$\quad h$ = thickness
$\quad \delta$ = total deflection
$\quad \sigma$ = nominal stress at built-in edge (stress concentration neglected)

Where the axial load P (Fig. 5.15) is not small compared with the buckling loads, Eqs. (5.92) and (5.93) no longer apply accurately. In such cases a more acccurate analysis shows that the stress and deflection may be found by multiplying the results calculated from these equations by factors C_1 and K_2, which depend on the ratio $P/P_{cr} = Pl^2/EI\pi^2$, where P_{cr} is the Euler buckling load for hinged ends. These factors are

$$C_1 = \frac{1}{1 - P/P_{cr}} \tag{5.94}$$

$$K_2 = 1 - 0.178 \frac{P}{P_{cr}} \tag{5.95}$$

The stress and deflection thus become

$$\sigma = K_2 \, \frac{3\delta Eh}{l^2} \tag{5.96}$$

$$\delta = C_1 \, \frac{Ql^3}{12EI} \tag{5.97}$$

The stress σ represents the range due to a lateral deflection δ.

A more exact expression for deflection δ of the spring of Fig. 5.15 is

$$\delta = \frac{Ql}{P} \, \frac{(2 \tan kl)/2 - kl}{kl} \tag{5.98}$$

where $$k = \sqrt{\frac{P}{EI}}$$

Leaf Springs

Leaf springs, while somewhat less efficient than helical springs in terms of energy storage per pound of material, are widely used in automotive applications, since they may function as structural members also.

The practical design of leaf springs is sometimes based on the assumption of a beam of uniform strength, which is equivalent to assuming a triangular-profile cantilever spring. On this assumption, rate R and maximum stress σ for a *symmetric* semielliptic leaf spring (Fig. 5.16a) are

$$R = \frac{P}{\delta} = \frac{8Enbh^2}{3l^3} \tag{5.99}$$

$$\sigma = \frac{3Pl}{2nbh^2} \tag{5.100}$$

where n = number of leaves
b = width
h = thickness of leaf
l = length
P = load

For an *unsymmetric* semielliptic spring (Fig. 5.16b),

$$\text{Rate } R = \frac{P}{\delta} = \frac{Ebnh^3l}{6l_1^2l_2^2} \tag{5.101}$$

$$\text{Stress } \sigma = \frac{6Pl_1l_2}{nbh^2l} \tag{5.102}$$

For the cantilever leaf spring (Fig. 5.16c) having n leaves, l being the length of the longest leaf,

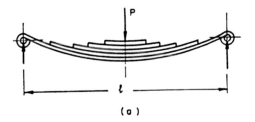

(a)

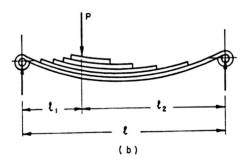

(b)

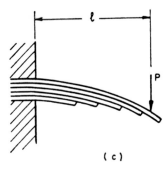

(c)

FIGURE 5.16   Leaf springs. (*a*) Symmetric semielliptic;
(*b*) unsymmetric semielliptic; (*c*) cantilever.

$$\text{Rate } R = \frac{P}{\delta} = \frac{Ebnh^3}{6l^3} \tag{5.103}$$

$$\text{Stress } \sigma = \frac{6Pl}{nbh^2} \tag{5.104}$$

TORSION-BAR SPRINGS

General

Torsion-bar springs consist essentially of straight bars of a spring material (Fig. 5.17) subject primarily to torsion. Energy is stored because of the twisting of the bar. Such springs are used for automotive vehicle suspensions as well as other applications.

For torsion-bar springs loaded in pure torsion, the following design formulas for various cross sections apply. (If bending moments act on the spring, additional bending stresses must be considered.)

Solid Round Bar

$$\phi = \frac{584M_t l}{d^4 G} \tag{5.105}$$

$$\tau = \frac{16M_t}{\pi d^3} \tag{5.106}$$

where ϕ = wind-up angle, deg
τ = shear stress
M_t = applied torque moment
d = bar diameter
l = active length of spring
G = modulus of rigidity

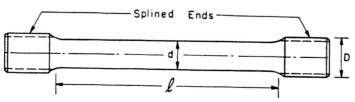

FIGURE 5.17   Torsion-bar spring with splined ends.

Hollow Round Bar

If the bar is hollow, with outside diameter d and inside diameter d_1, the formulas become

$$\phi = \frac{584 M_t l}{G(d^4 - d_1^4)} \tag{5.107}$$

$$\tau = \frac{16 M_t d}{G(d^4 - d_1^4)} \tag{5.108}$$

Square Bar

$$\phi = \frac{407 M_t l}{a^4 G} \tag{5.109}$$

$$\tau = \frac{4.81 M_t}{a^3} \tag{5.110}$$

where a = side of square cross section.

Rectangular Bar

$$\phi = \frac{57.3 M_t l}{k_1' a h^3 G} \tag{5.111}$$

$$\tau = \frac{M_t}{k_2' a h^2} \tag{5.112}$$

where a = length and h = thickness of rectangular cross section, and k_1' and k_2' depend on a/h.

Torsion Spring Loaded by Lever

Frequently torsion springs are loaded by a lever AB (Fig. 5.18) attached to one end, which is supported by a bearing; the other end is fixed. If the deflection δ is measured from the horizontal and the load P is vertical, then for a round bar

$$P = \frac{\pi d^4 G(\alpha + \beta)}{32 l R \cos \alpha} \tag{5.113}$$

Here α and β are angles measured in radians, with α = angle between lever

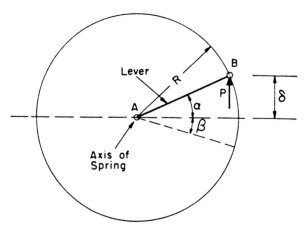

FIGURE 5.18 Torsion-bar spring loaded by lever.

centerline AB and horizontal reference line and β = angle between lever centerline and reference line at zero load.

The vertical rate at the end B of the lever (Fig. 5.18) is defined as $dP/d\delta = k$.

$$k = \frac{\pi d^4 G}{32 l R^2} \frac{1 + (\alpha + \beta) \tan \alpha}{\cos^2 \alpha} \tag{5.114}$$

RUBBER SPRINGS AND MOUNTINGS

The deflection δ of an unbonded compression block (Fig. 5.19) may be estimated approximately from the following formula, provided that com-

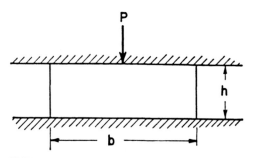

FIGURE 5.19 Loaded compression block of rubber.

pression surfaces are well lubricated so that lateral expansion may occur freely and that small strains are assumed.

$$\delta = \frac{Ph/AE}{1 + P/AE} \tag{5.115}$$

where P = load
A = original cross-sectional area of sandwich
h = original thickness

Equation (5.15) is based on constant volume as the block is compressed.

SHEAR SPRINGS OR SANDWICHES

Rubber shear springs or sandwiches which consist essentially of two rubber pads bonded to steel plates (Fig. 5.20) are widely used for vibration isolation and machine mounting. The shear stress $\tau = P/2A$, where A is the sectional area of each pad. The shear angle γ is equal to τ/G, where G = modulus of rigidity; and for small deflections $\gamma = \delta/h$ rad where δ = deflection and h = thickness of pad. This gives

$$\delta = \frac{Ph}{2AG} \tag{5.116}$$

Values of modulus G for small deflections are given as functions of durometer hardness in engineering handbooks.

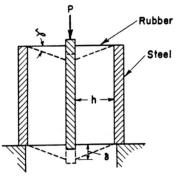

FIGURE 5.20 Simple shear spring or sandwich.

CYLINDRICAL SHEAR SPRING

Constant Axial Height

This type of shear spring consists essentially of a circular pad bonded to a steel ring on the outside and to a shaft or ring on the inside (Fig. 5.21). A load P is applied along the axis. The shear stress τ at any radius r will be

$$\tau = \frac{P}{2\pi rh} \tag{5.117}$$

If y = deflection at radius r, $dy/dr = -\tau/G$ approximately. By using the toal deflection, δ becomes approximately

$$\delta \approx \frac{P}{2\pi hG} \ln \frac{r_o}{r_i} \tag{5.118}$$

Constant Stress

If the thickness h of a cylindrical rubber spring is made inversely proportional to the radius r (Fig. 5.22), the shear stress τ will be constant, and better utilization of the material is obtained. By taking $r = r_o h_o / h$, where h_o = thickness at the outer radius r_o,

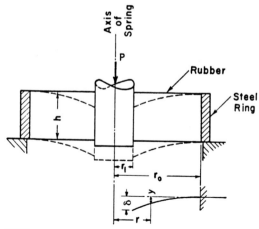

FIGURE 5.21 Cylindrical shear spring of constant axial height with axial load.

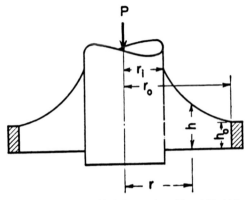

FIGURE 5.22 Cylindrical shear spring with variable thickness and constant shear stress.

$$\tau = \frac{P}{2\pi r_o h_o} = \text{const} \tag{5.119}$$

The deflection δ will be approximately equal to τ/G multiplied by $r_o - r_i$.

$$\delta = \frac{P(r_o - r_i)}{2\pi r_o h_o G} \tag{5.120}$$

CYLINDRICAL TORSION SPRING

Constant Thickness

In this case the thickness h of the spring is taken as constant, (Fig. 5.23) while a moment M is assumed to act about the spring axis. The shear stress τ at radius r due to moment M is

$$\tau = \frac{M}{2\pi r^2 h} \tag{5.121}$$

In this case, the maximum shear stress τ_m will occur when $r = r_i$ and is

$$\tau_m = \frac{M}{2\pi r_i^2 h} \tag{5.122}$$

Angular deflection is

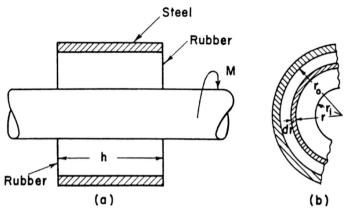

FIGURE 5.23 Cylindrical rubber torsion spring of constant thickness.

$$\theta = \frac{M}{4\pi hG} \left(\frac{1}{r_i^2} - \frac{1}{r_o^2} \right) \tag{5.123}$$

where θ is given in radians.

FLAT BELTS

Belt Lengths

The total belt length L (Fig. 5.24) is

$$L = 2S \cos \theta + \pi[R + r + (R - r)\theta/90] \qquad \text{in} \qquad (5.124)$$

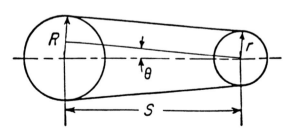

FIGURE 5.24 Typical belt pulley drive.

where S = center distance between pulleys
  R = radius of large pulley (flat belt) or pitch
        radius of large pulley (V belt)
  r = radius of small pulley (flat belt) or pitch
        radius of small pulley (V belt)
  θ = $\sin^{-1} [(R - r)/S]$, deg

The velocity ratio is the ratio of the angular velocity of the driving shaft to the angular velocity of the driven shaft. Thus

$$\frac{N_1}{N_2} = \frac{D_2}{D_1} \tag{5.125}$$

and
$$N_2 D_1 = N_2 D_2 \tag{5.126}$$

where N_1 = rotation speed of driving shaft, rpm
  N_2 = rotation speed of driven shaft, rpm
  D_1 = diameter of driving pulley
  D_2 = diameter of driven pulley

For a flat belt, the thickness of the belt is ignored, and the diameter of the pulley is the outside diameter.

For a chain drive, the following relation must be used

$$\frac{N_1}{N_2} = \frac{n_2}{n_1} \tag{5.127}$$

where n_1 = number of teeth on driving sprocket and n_2 = number of teeth on driven sprocket.

Belt Velocity

The velocity at which a belt travels may be found from

$$V = \frac{\pi D N}{12} = 0.262 D N \tag{5.128}$$

where V = velocity, ft/min (m/min)
  D = pulley diameter, in (cm)
  N = rpm

Belt Tension

Referring to Fig. 5.25, where F_1 is the tension in the tight side and F_2 is the tension in the slack side, we see that the torque at the small pulley is

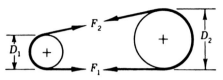

FIGURE 5.25 Forces in a belt.

$$T_1 = (F_1 - F_2)\frac{D_1}{2} \tag{5.129}$$

And the torque at the large pulley is

$$T_2 = (F_1 - F_2)\frac{D_2}{2}$$

The term $F_1 - F_2$ is known as the *net tension.*

Power Delivered

The equation for the horsepower transmitted by a belt is

$$\text{hp} = F_e \frac{0.262DNW}{33,000} = \frac{F_e DNW}{126,000} \tag{5.130}$$

where F_e = net tension, lb/in (kg/cm), of belt width
W = belt width, in (cm)

ROLLER CHAINS

The average velocity of a chain is

$$V = \frac{pNn}{12} \tag{5.131}$$

where V = average velocity, ft/min (kg)
p = chain pitch, in (cm)
N = rpm of sprocket
n = number of teeth on sprocket

The approximate length of a chain can be found by

$$L = \frac{n_1 + n_2}{2} + \frac{2C}{p} + \frac{p(n_1 + n_2)^2}{39.5C} \tag{5.132}$$

where L = length of chain in links
n_1, n_2 = numbers of sprocket teeth
C = center distance, in (cm)
p = chain pitch, in (cm)

The equation for the horsepower hp transmitted by a chain is

$$\text{hp} = \frac{F_a V}{33,000} \tag{5.133}$$

where F_a = allowable tension in chain, lb (kg), and V = velocity of chain, ft/min (m/min). Neglecting the centrifugal force, which at the recommended velocity is negligible, we see that the commonly used equation for the allowable tension in the chain is

$$F_a = \frac{2,600,000A}{V + 600} \tag{5.134}$$

where A = projected area of the pin joint, in^2 (cm^2). For standard chain, all the dimensions are relative to the pitch, and the projected area of the pin joint is

$$A = 0.273p^2 \tag{5.135}$$

Standard pitch sizes are: ¼, ⅜, ½, ⅝, ¾, 1, 1¼, 1½, 1¾, 2, and 2½ in.

SILENT CHAINS

The horsepower transmitted is

$$\text{hp} = \frac{T_p VW}{33,000} \tag{5.136}$$

where T_p = permissible tension, lb/in (m/min), and W = width of chain, in (cm). The total capacity of the chain should be greater than the normal power to be transmitted. The amount is dictated by the application. For severe shock, an appropriate chain capacity would be twice the normal power.

BLOCK BRAKES

A block brake in its simplest form is merely a shaped block pressed against the shaft, as shown in Fig. 5.26. The force transmitted to the block by friction, known as the friction force, is

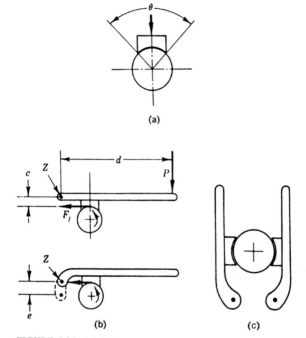

FIGURE 5.26 (a) A simple block brake; (b) block brake arrangements; (c) a double block brake.

$$F_f = fF_n \tag{5.137}$$

where F_f = friction force, lb (N)
  f = coefficient of friction
  F_n = load forcing block against drum, lb (N)

This force may also be thought of as the resistance that the block offers to the rotation of the drum. Thus, the resisting torque is

$$T = F_f r \quad \text{or} \quad T = fF_n r \tag{5.138}$$

where T = resisting torque, lb · in (N · m), and r = radius of drum, in (mm). Equation (5.138) may be used for blocks of a length such that the angle of contact (θ in Fig. 5.26a) is less than 60°. For larger angles where the pressure on the block is unevenly distributed, use

$$T = \frac{4fF_n r \sin(\theta/2)}{\theta + \sin \theta} \tag{5.139}$$

where θ = angle of contact, rad.

BAND BRAKES

A band brake consists of a steel band lined on the inside with a friction material, as shown in Fig. 5.27. The operating force is applied so as to tighten the band around the drum. The difference between tensions F_1 and F_2 is the friction force in the same manner as the difference between tensions was the driving force of a belt. The braking torque then is

$$T = r(F_1 - F_2) \tag{5.140}$$

where r = drum radius, in (mm). The tensions can be found with

$$\frac{F_1}{F_2} = e^{\theta f} \tag{5.141}$$

and
$$F_1 = rbp_a$$

where e = mathematical constant 2.718
$\quad \theta$ = angle of contact, rad
$\quad b$ = band width, in (mm)
$\quad p_a$ = maximum allowable pressure, psi (MPa)

The force required to apply the brake is

$$P = F_2 \frac{a}{a + b} \tag{5.142}$$

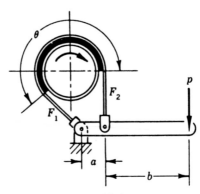

FIGURE 5.27 A band brake.

CLUTCHES

Plate Clutches

A simple plate clutch is shown in Fig. 5.28. Part A is secured to its shaft, part B is keyed to its shaft with a feather key, and the facing C is attached to B. Part B is forced against A to engage the clutch and away from A to disengage. And D and d are the two diameters of the facing. Because the facing close to the axis of rotation is ineffective, d is seldom less than ½ D.

The design of a clutch is based on the assumption that either the pressure is uniformly distributed or the axial wear is uniform, neither of which is actually the case. However, an equation that has proved successful is

$$T = \frac{fP(D + d)}{4} \tag{5.143}$$

where T = torque transmitted, lb · in $(N \cdot m)$
$\ P$ = axial load, lb (N)
$\ f$ = coefficient of friction

Cone Clutches

The cone clutch shown in Fig. 5.29 is a simple friction clutch which has the advantage that a small axial load will produce a large force pressing the friction surfaces together. The angle α should not be less than 8° and is usually not more than 15°. The torque transmitted can be found with Eq. (5.143) when modified to allow for the angle as follows:

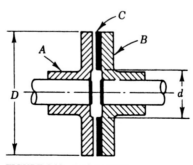

FIGURE 5.28   A plate clutch.

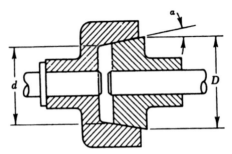

FIGURE 5.29 A cone clutch.

$$T = \frac{fP(D + d)}{4 \sin \alpha} \qquad (5.144)$$

where f = the coefficient of friction and the other symbols are as shown in Fig. 5.29.

KEYWAYS

The torque on a keyway (Fig. 5.30) is given by

$$T = WLs_s \times \frac{D}{2} \qquad (5.145)$$

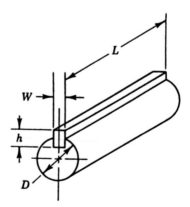

FIGURE 5.30 Key dimensions.

where T = torque, lb · in (N · m)
  L = length of key, in (cm)
  W = width of key, in (cm)
  h = height of key, in (cm)
  D = diameter of shaft, in (cm)
  s_s = shear stress, psi (cm)
  s_c = compression stress, psi (MPa)

Also,

$$T = \frac{h}{2} \times Ls_c \times \frac{D}{2} \tag{5.146}$$

Hence,

$$s_c = \frac{4T}{hLD} \tag{5.147}$$

Keys are often designed to fail before the shaft or hub fails because it is easier and less costly to replace the key. The size h of the key should be about one-fourth the shaft diameter D.

GEARS

Points on the two cylinders in Fig. 5.31 have the same linear velocity if it is assumed that there is no slipping. It follows, therefore, that

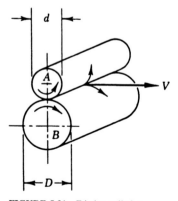

FIGURE 5.31 Friction cylinders.

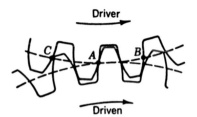

FIGURE 5.32   Gear teeth in mesh.

$$\text{rpm of } Ad = \text{rpm of } BD \qquad (5.148)$$

To prevent slipping, teeth may be so arranged on each cylinder that the effective diameters are not changed, as shown in Fig. 5.32 where the dashed lines are equivalent to the surfaces of the cylinders in Fig. 5.31. Adding the teeth does not change Eq. (5.148).

Spur Gear Formulas

For gears with either a 14.5° or 20° pressure angle (Fig. 5.33) the terms and formulas below apply, with dimensions in inches or centimeters.

Pitch circle: An imaginary circle that corresponds to the dashed lines in Fig. 5.34.

Pitch diameter D: The diameter of the pitch circle. The diameter of a gear is understood to be the pitch diameter.

Number of teeth n: The number of teeth on the gear.

Diametral pitch P: The number of teeth on the gear per inch of pitch diameter. $P = n/D$. The pitch of a gear is understood to be the diametral pitch.

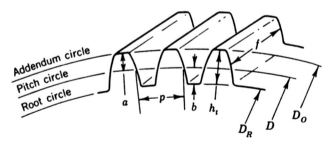

FIGURE 5.33   Spur gear terminology.

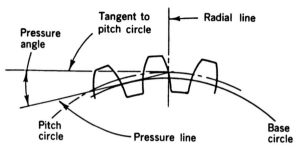

FIGURE 5.34 Pressure angle.

Addendum a: The radial distance from the pitch circle to the addendum circle. $a = 1/P$.

Dedendum b: The radial distance from the pitch circle to the bottom of the tooth space. $b = 1.157/P$.

Outside diameter D_o: The diameter of the addendum circle. $D_o = D + 2a$; also $D_o = (n + 2)/P$.

Root diameter: D_R: The diameter of the root circle. $D_R = D - 2b$. Whole depth: h_t: The total height. $h_t = a + b$.

Whole depth h_t: The total height. $h_t = a + b$.

Face width f: The width of the gear tooth.

Circular pitch p: The distance measured along the pitch circle from a point on one tooth to the corresponding point on the adjacent tooth. $p = \pi D/n$. As $P = n/D$, $\pi = pP$.

Working depth h_k: The distance that a tooth on one gear projects into the space between the teeth on the other gear. $h_k = 2a$.

Circular thickness t_c: The thickness of a tooth measured along the pitch circle $t_c = p/2$.

Power-Velocity Load

Power is transmitted by one gear exerting a force on the other. This force

$$F = \frac{33,000 \text{ hp}}{V} \qquad (5.149)$$

where F = force on gear tooth, lb (kg)
   hp = horsepower transmitted
   V = velocity at pitch circle, ft/min (m/min)

The velocity at pitch circle is

$$V = 0.262DN \qquad (5.150)$$

where N = rotation speed, rpm.

Gear Strength

The capacity of a gear to transmit power is dependent on the strength of a tooth acting as a cantilever beam. The allowable load is

$$F_s = \frac{sfY}{P} \qquad (5.151)$$

where F_s = allowable load, lb (kg)
  s = allowable stress, psi (MPa)
  Y = tooth form factor

The allowable stress and form factor can be obtained from charts and tables in engineering handbooks.

To allow for shock loads and manufacturing imperfections, the allowable load found with Eq. (5.151) should be multiplied by a factor K. The value of K for commercial-quality gears at less than 2000 ft/min (610 m/min) is

$$K = \frac{600}{600 + V} \qquad (5.152)$$

For accurately hobbed or generated gears at velocities less than 4000 ft/min (1219 m/min), the value is found with

$$K = \frac{1200}{1200 + V} \qquad (5.153)$$

For precision gears that are ground or lapped and operating at velocities over 4000 ft/min (1219 m/min), K is found with

$$K = \frac{78}{78 + \sqrt{V}} \qquad (5.154)$$

Backlash

For most gear applications the backlash clearance is about $0.04/P$. The force tending to separate two meshing gears carrying a load is

$$S = F \tan \alpha \qquad (5.155)$$

where S = separating force, lb (N), and α = tooth pressure angle.

ACME SCREWS

The torque required to exert a given force with an Acme screw is

$$T = \frac{Qd}{2}\left(\frac{\cos\alpha\,\tan\lambda + \mu}{\cos\alpha - \mu\,\tan\lambda}\right) \tag{5.156}$$

where T = torque, lb·in (N·m)
Q = load or force, lb (N)
d = pitch diameter, in (mm)
λ = lead angle
μ = coefficient of friction

The torque required to overcome the friction at the thrust collar is found with

$$T_c = \frac{\mu Q d_c}{2} \tag{5.157}$$

where T_c = torque, lb·in (N·m), and d_c = mean diameter of thrust collar, in (mm). The coefficient of friction for a steel collar on a bronze thrust bearing is about 0.1 starting and 0.08 running. If a ball or roller thrust bearing is used, the friction is so small relative to that of the screw that it may be ignored.

The total torque required to exert the force Q is the sum of the torque from Eq. (5.156) and the torque from Eq. (5.157).

COLUMNS IN MACHINE PARTS

Column action due to axial loading of machine parts occurs very frequently. If the axial load is a tensile load, then the application of $S = P/A$ is in order. If the axial load is a compressive load, then an appropriate column equation should be used.

The Euler equation for the critical load for slender columns of uniform cross section is

$$F_{cr} = \frac{C\pi^2 EA}{(L/k)^2} \tag{5.158}$$

where F_{cr} = critical load to cause buckling
C = constant depending upon end conditions (see engineering handbooks for values)
E = modulus of elasticity, psi
A = area of transverse section, in^2
L = length of column, in

k = minimum radius of gyration = $\sqrt{I/A}$ in, where I is the minimum moment of inertia about the axis of bending. For a circular section, $k = D/4$. For a rectangular section $k = h\sqrt{3}/6$, where h is the smaller dimension of the rectangle.

The critical load for moderate-length columns of uniform cross section is given by several empirical formulas, one of which is the J. B. Johnson formula:

$$F_{cr} = s_y A \left[1 - \frac{s_y(L/k)^2}{4C\pi^2 E} \right] \tag{5.159}$$

where s_y = yield point, psi. The other symbols are as defined for the Euler equation. The value of C depends on the end conditions (Fig. 5.35).

The safe load is obtained by dividing the critical load by a factor of safety N:

Safe load F, Euler equation: $F = \dfrac{F_{cr}}{N} = \dfrac{C\pi^2 EA}{N(L/k)^2}$ (5.160)

Safe load F, Johnson equation: $F = \dfrac{s_y A}{N} \left[1 - \dfrac{s_y(L/k)^2}{4C\pi^2 E} \right]$ (5.161)

One end fixed and the other end free of all restraint

$C = \frac{1}{4}$

Both ends free to rotate, but not free to move laterally (so-called round, or pivot, or hinged end columns)

$C = 1$

One end fixed and one end free to rotate, but not free to move laterally

$C = 2$

Both ends fixed so that the tangent to the elastic curve at each end is parallel to the original axis of the column

$C = 4$

FIGURE 5.35

TABLE 5.1

C	E, psi	s_y, psi	$(L/k)^2$	L/k
¼	30×10^6	80,000	1,849	43
		70,000	2,113	46
		60,000	2,465	50
		50,000	2,958	54
		40,000	3,697	61
1	30×10^6	80,000	7,394	86
		70,000	8,451	92
		60,000	9,860	99
		50,000	11,832	109
		40,000	14,789	121
2	30×10^6	80,000	14,789	121
		70,000	16,902	130
		60,000	19,719	140
		50,000	23,663	154
		40,000	29,579	172

If L/k is below that given by $\sqrt{2C\pi^2E/s_y}$, use the Johnson formula, which is valid down to $L/k = 0$.

The value of L/k which determines whether the Euler equation or the Johnson equation should be used is found by equating the critical load from the Euler equation to the critical load from the Johnson formula:

$$\frac{C\pi^2EA}{(L/k)^2} = s_yA\left[1 - \frac{s_y(L/k)^2}{4C\pi^2E}\right] \tag{5.162}$$

from which
$$L/k = \sqrt{\frac{2C\pi^2E}{s_y}}$$

The values of L/k above which the Euler equation should be used and below which the Johnson formula should be used, for different representative data, are as shown in Table 5.1.

RELIABILITY OF MACHINE COMPONENTS AND SYSTEMS

Reliability is the characteristic of a component, or of a system made up of many components, expressed by the probability that it will perform its particular function within a specific environment for a given time. Reliability predictions have become a precise branch of industrial technology. Reliability engineering plays an important role in the reduction of costly failures and correct planning of overhaul and maintenance schedules.

Summary of Relevant Formulas

For a constant proportional failure rate,

$$R = e^{-\lambda t} \tag{5.163}$$

$$R + Q = 1 \tag{5.164}$$

$$Q = 1 - e^{-\lambda t} \tag{5.165}$$

$$N_s = N_0 e^{-\lambda t} \tag{5.166}$$

$$N_f = N_0(1 - e^{-\lambda t}) \tag{5.167}$$

$$m = \frac{1}{\lambda} \tag{5.168}$$

$$R = e^{-t/m} \tag{5.169}$$

$$Q = 1 - e^{-t/m} \tag{5.170}$$

$$N_s = N_0 e^{-t/m} \tag{5.171}$$

$$N_f = N_0(1 - e^{-t/m}) \tag{5.172}$$

where R = reliability
Q = unreliability
λ = proportional failure rate (i.e., failure rate expressed as a proportion of N_0)
N_s = number of live components (survivors)
N_f = number of dead components (failures)
N_0 = initial number of live components
m = mean time between (chance) failures (MTBF)
t = time

For a test to determine MTBF m,

$$m = \frac{\text{total survival hours}}{\text{number of failures}} \tag{5.173}$$

For a normally distributed variable x, the standard deviation σ is given by

$$\sigma = \sqrt{\frac{\Sigma(x - x_m)^2}{n}} \tag{5.174}$$

where x_m is the mean value of n observations of x.

In a test to determine mean wearout life M,

$$M = \frac{\text{sum of lives}}{\text{number of components}} \qquad (5.175)$$

For an exponentially distributed variable,

$$\text{Upper confidence limit} = \frac{2nm}{\chi^2_{1-\alpha/2,n}} \qquad (5.176)$$

$$\text{Lower confidence limit} = \frac{2nm}{\chi^2_{\alpha/2,n}} \qquad (5.177)$$

at a level of confidence given by $100(1 - \alpha)$ percent, where n denotes number of failures, m denotes an estimate of the mean value of the variable, and χ^2 denotes the value of chi squared (given in tables) for values of n and α or $1 - \alpha/2$.

For a component or unit that forms part of a system

$$\frac{m_c}{m_s} = \frac{t_c}{t_s} \qquad (5.178)$$

$$m_s = \frac{m_c}{d} \qquad (5.179)$$

where m_c = component MTBF in component operating hours
m_s = component MTBF in system operating hours
t_c = component operating hours
t_s = system operating hours
d = duty cycle = t_c/t_s

The probability of both events x and y occurring, denoted by P_{xy}, is given by

$$P_{xy} = P_x P_y \qquad (5.180)$$

And the probability of either event x or event y occurring, denoted by P_{x+y}, is given by

$$P_{x+y} + P_x + P_y - P_x P_y \qquad (5.181)$$

where P_x = the probability of x occurring and P_y = the probability of y occurring.

The following equations refer to series and parallel systems. The symbols used are as above for reliability, unreliability, failure rate, etc., with the addition of appropriate subscripts as follows:

Subscript s denotes series

Subscript p denotes parallel

Subscripts 1, 2, 3, etc., denote components or subunits 1, 2, 3, etc.

Figure 5.36 shows reliability curves for series and parallel systems.

$$R_s = R_1 R_2 \cdots \tag{5.182}$$

$$Q_s = Q_1 + Q_2 - Q_1 Q_2 \tag{5.183}$$

$$Q_s = 1 - R_s \tag{5.184}$$

$$R_p = R_1 + R_2 - R_1 R_2 \tag{5.185}$$

$$Q_p = Q_1 Q_2 \tag{5.186}$$

$$Q_p = 1 - R_p \tag{5.187}$$

$$R_s = e^{-(\lambda_1 + \lambda_2 + \lambda_3 + \cdots)t}. \tag{5.188}$$

$$\lambda_s = \lambda_1 + \lambda_2 + \lambda_3 + \cdots \tag{5.189}$$

$$m_s = \frac{1}{\lambda_s} \tag{5.190}$$

$$R_s = e^{-n\lambda t} \tag{5.191}$$

$$\lambda_s = n\lambda \tag{5.192}$$

$$m_s = \frac{1}{n\lambda} \tag{5.193}$$

$$R_p = R_1 + R_2 - R_1 R_2 \tag{5.194}$$

$$R_p = R_1 + R_2 + R_3 - R_1 R_2 - R_2 R_3 - R_1 R_3 + R_1 R_2 R_3 \tag{5.195}$$

$$R_p = e^{-\lambda_1 t} + e^{-\lambda_2 t} + e^{-(\lambda_1 + \lambda_2)t} \tag{5.196}$$

$$R_p = e^{-\lambda_1 t} + e^{-\lambda_2 t} + e^{-\lambda_3 t} + e^{-(\lambda_1 + \lambda_2)t} + e^{-(\lambda_2 + \lambda_3)t}$$
$$+ e^{-(\lambda_1 + \lambda_3)t} + e^{-(\lambda_1 + \lambda_2 + \lambda_3)t} \tag{5.197}$$

$$m_p = \frac{1}{\lambda_1} + \frac{1}{\lambda_2} - \frac{1}{\lambda_1 + \lambda_2} \tag{5.198}$$

$$m_p = \frac{1}{\lambda_1} + \frac{1}{\lambda_3} + \frac{1}{\lambda_3} - \frac{1}{\lambda_1 + \lambda_2} - \frac{1}{\lambda_2 + \lambda_3}$$
$$- \frac{1}{\lambda_1 + \lambda_3} + \frac{1}{\lambda_1 + \lambda_2 + \lambda_3} \tag{5.199}$$

$$m_p = \frac{1}{\lambda} + \frac{1}{2\lambda} + \frac{1}{3\lambda} + \cdots + \frac{1}{n\lambda} \tag{5.200}$$

where n in Eqs. (5.192), (5.193), and (5.200) denotes the number of components or subunits having equal failure rates.

For a system, the utilization factor U is given by

$$U = \frac{\text{operating time}}{\text{maintenance time} + \text{idle time} + \text{operating time}} \tag{5.201}$$

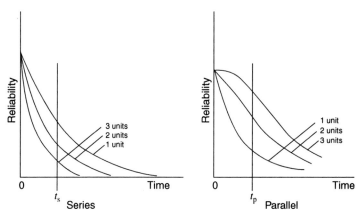

FIGURE 5.36 Reliability curves for series and parallel systems.

The availability (maximum utilization factor) A is given by

$$A = U_{max}$$

$$= \frac{\text{operating time}}{\text{minimum maintenance time + operating time}} \qquad (5.202)$$

For any two sets of operating conditions denoted by x and m, respectively, the voltages V_x and V_m, temperatures t_x and t_m, and failures rates λ_x and λ_m are related by the equation

$$\lambda_x = \lambda_m \left(\frac{V_x}{V_m}\right)^n K^{t_x - t_m} \qquad (5.203)$$

where n and K are constants over a limited range of conditions and may be determined by the equations

$$K = \frac{\lambda_x}{\lambda_m} \frac{1}{t_x - t_m} \qquad (5.204)$$

for a constant-voltage test and

$$n = \frac{\ln (\lambda_x / \lambda_m)}{\ln (V_m / V_x)} \qquad (5.205)$$

for a constant-temperature test.

GENEVA WHEEL DESIGN

The Geneva wheel (Fig. 5.37) was originally used as a stop for preventing overwind in watch springs—leave one of the slot positions unslotted, and the number of turns the drive can make is limited. Now, the Geneva wheel

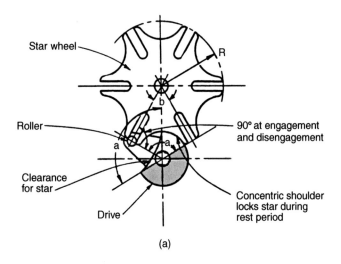

(a)

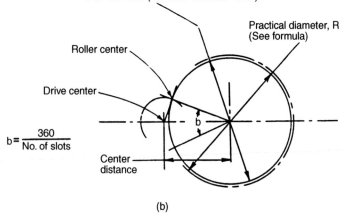

$$b = \frac{360}{\text{No. of slots}}$$

(b)

FIGURE 5.37 Geneva wheel components. (*a*) Complete Geneva mechanism; (*b*) basic step in layout.

is one of the most useful mechanisms for providing high-speed, intermittent, rotary motion.

Design Formulas

Values such as angular position at any instant, velocity, acceleration, and practical star-wheel diameter can be calculated from the formulas that follow. Illustrated in Fig. 5.38 are symbols that appear in the formulas.

Center distance $A = CM$

where
$$M = \frac{1}{\text{Sin}(180/\text{no. of slots})} \qquad (5.206)$$

Angular displacement:

$$\text{Tan } \beta = \frac{\text{Sin } \alpha}{M - \cos \alpha} \qquad (5.207)$$

Angular velocity:
$$= \frac{d\beta}{dt}$$
$$= \omega \frac{M \cos \alpha - 1}{1 + M^2 - 2M \cos \alpha} \qquad (5.208)$$

Angular acceleration, radians/sec^2 rad/s^2

$$= \frac{d^2\beta}{dt^2}$$

$$= \omega^2 \frac{M \sin \alpha (1 - M^2)}{(1 + M^2 - 2M \cos \alpha)^2} \qquad (5.209)$$

Maximum acceleration occurs when

$$\cos \alpha = \pm \sqrt{\left(\frac{1 + M^2}{4M}\right)^2 + 2} - \frac{1 + M^2}{4M} \qquad (5.210)$$

Maximum acceleration and therefore maximum wear occur at about one-third to one-fourth the distance down the slot length from the wheel edge.
  Practical star-wheel diameter is

$$2R = \sqrt{A^2 - C^2 + \mu^2} \qquad (5.211)$$

STRESS FORMULAS FOR THICK CYLINDERS

When the ratio of the outer radius to the inner radius exceeds 1.14 to 1, a cylinder is usually classified as *thick*. The distribution of stresses is non-

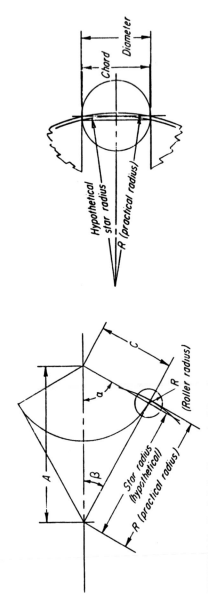

FIGURE 5.38 Symbols for Geneva wheel formulas.

TABLE 5.2 Comparison of Formulas

Author	Equation		Assumptions
Lamé	$S_t = a + \dfrac{b}{r^2} = p\left(\dfrac{1+R^2}{R^2-1}\right) = p\left(\dfrac{r_o^2+r_i^2}{r_o^2-r_i^2}\right)$	(5.212)	Brittle materials, closed ends
Barlow	$S_t = \dfrac{PD}{2t} = P\left(\dfrac{r_o}{t}\right) = p\left(\dfrac{R}{R-1}\right)$	(5.213)	Open ends
Clavarino	$S_t = (1-2m)a + \dfrac{1+m}{r_i^2}\,b$ $= \dfrac{0.4}{R^2-1} + \dfrac{1.3R^2}{R^2-1} = p\left(\dfrac{0.4+1.3R^2}{R^2-1}\right)$	(5.214)	Ductile material, closed ends
Birnie	$S_t = (1-m)a + (1+m)\dfrac{b}{r_i^2}$ $= p\left(\dfrac{0.7}{R^2-1} + \dfrac{1.3R^2}{R^2-1}\right) = p\left(\dfrac{0.7+1.3R^2}{R^2-1}\right)$	(5.215)	Ductile material, open ends

Nomenclature:
r_o = outside radius, in
r_i = inside radius, in
D = outside diameter, in
S_t = working stress, tension at inner radius, psi
p = working pressure, psi
m = Poisson's ratio (0.3 for steel)
t = wall thickness, $r_o - r_i$, in

$R = \dfrac{r_o}{r_i}$ $a = p\left(\dfrac{r_i^2}{r_o^2-r_i^2}\right)$ $b = p\left(\dfrac{r_i^2 r_o^2}{r_o^2-r_i^2}\right)$

127

TABLE 5.3 Formulas for Curved Springs

Spring type	Spring deflection	Spring force and bending stresses
A	$F_1 = \dfrac{KPr^3}{3EI}(m + \beta)^3$ where $\alpha = \beta$ for finding K	When $a = 0°$ to $90°$ $P = \dfrac{s\sigma}{u + \sin\beta}$ $\sigma = \dfrac{Pr(m + \sin\beta)}{S}$ When $a = 90°$ to $180°$ $P = \dfrac{S\sigma}{u + r}$ $\sigma = \dfrac{Pr(m + 1)}{S}$
B	$F_2 = \dfrac{2KPr^3}{3EI}\left(m + \dfrac{\beta}{2}\right)^3$ where $a = \beta/2$ for finding K	$P = \dfrac{S\sigma}{L}$ $\sigma = \dfrac{Pl}{S}$
	$F_3 = 2F_2 = \dfrac{4KPr^3}{3EI}\left(m + \dfrac{\beta}{2}\right)^3$ where $\alpha = \beta/2$ for finding K	

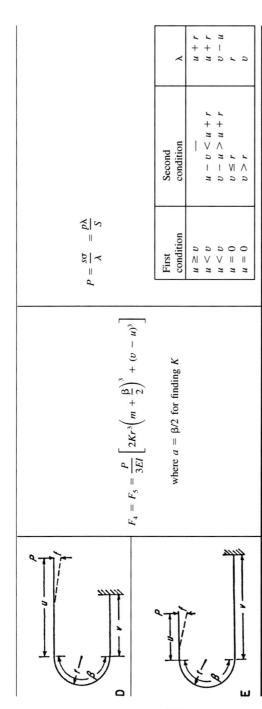

$$P = \frac{s\sigma}{\lambda} = \frac{p\lambda}{S}$$

First condition	Second condition	λ
$u \geq v$	—	$u + r$
$u < v$	$u - v < u + r$	$u + r$
$u < v$	$v - u > u + r$	$v - u$
$u = 0$	$v \leq r$	r
$u = 0$	$v > r$	v

$$F_4 = F_5 = \frac{P}{3EI}\left[2Kr^3\left(m + \frac{\beta}{2}\right)^3 + (v - u)^3 \right]$$

where $a = \beta/2$ for finding K

D

E

Nomenclature:

b = spring width, mm; D = spring diameter, mm; E = Young's modulus, kg/mm^2; F = spring displacement, mm; h = spring thickness, mm; I = moment of inertia, mm^4; K = correction factor; L = moment arm, mm; $m = u/r$; P = spring force, kg; r = radius of curvature, kg; S = section modulus, mm^3; u = length of straight section at free end of spring, mm; v = length of straight section at fixed end of spring, mm; α = parameter for plotting K; β = angle of spring curvature, rad; λ = moment arm, mm; σ = maximum bending stress, kg/mm^3.

For accurate results h/r for flat springs or D/r for round wire springs should be less than 0.6.

129

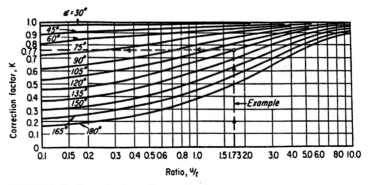

FIGURE 5.39   Correction factors for curved spring.

uniform radially and a maximum at the inner radius. This stress variation
is verified by mathematical analysis as well as actual tests. To compute the
stress values at the inner wall for different end conditions and materials,
several formulas are in use, each identified by its authors. These formulas
are given in Table 5.2 for four different recognized and respected authors.

DESIGN FORMULAS FOR CURVED SPRINGS

Curved springs are usually designed with formulas based on the stretched
length of the spring, disregarding curvature influence. These equations are
inaccurate. The new method retains simplicity but gains accuracy by using
a correction factor for curvature (Table 5.3 and Fig. 5.39). Test results agree
closely with deflections as calculated by these formulas; for example, a type
D spring was calculated within 8 percent of its actual deflection and stress
compared to more than 100 percent error when formulas that ignore curvature were used. Curved spring shapes are tabulated below as five basic
types—com-
plex spring shapes can be computed by splitting them up into their basic
shapes.

   These formulas and the correction factors are the work of Joachim Palm
and Klaus Thomas of Munich, Germany.

HYDRODYNAMIC FORMULAS USEFUL IN
BEARING DESIGN

Bearing design requires a number of hydrodynamic formulas involving hydraulics, fluid flow, power, pressure head, torque, fluid viscosity, and fluid
density. The formulas in Table 5.4 present useful data in both USCS and
SI units.

TABLE 5.4 Hydrodynamic Formulas Useful in Bearing Design

Name	Unit	Symbol	Formula or value	System
Mass	Slugs	M	$\dfrac{\text{lb} \times \text{s}^2}{\text{ft}}$	USCS
Kilogram mass	Metric slug	M	$\dfrac{\text{kg} \times \text{s}^2}{\text{m}}$	SI
Gram mass	$\dfrac{\text{dyn} \times \text{s}^2}{\text{cm}}$	M	$\dfrac{\text{dyn} \times \text{s}^2}{\text{cm}}$	SI
Gravitational constant		g	32.174 (in London)	USCS
		g	9.807 (in Paris)	SI
Force	dyn	P	$\dfrac{1}{981}\,g$	SI
	Poundal	P	$\dfrac{1}{32.174}\,\text{lb}$	USCS
Pressure head	ft	H	For water, 1 ft equals 0.433 lb/in^2 or 62.335 lb/ft^2	USCS
Rated work	hp	N	550 ft · lb/s or 33,000 ft · lb/min	USCS

131

TABLE 5.4 Hydrodynamic Formulas Useful in Bearing Design (*Continued*)

Name	Unit	Symbol	Formula or value	System
Rated work	hp	N	$\dfrac{ft^3}{s} \times \dfrac{head \times sp\ gr}{8.8}$ or	USCS
			$\dfrac{gal}{min} \times \dfrac{head \times sp\ gr}{3960}$	
	hp	N	$\dfrac{ft^3}{min} \times \dfrac{lb}{ft^2} \times \dfrac{1}{33{,}000}$ or	USCS
			$\dfrac{gal}{min} \times \dfrac{lb}{in^2} \times \dfrac{1}{1714}$	
Torque	lb · ft	T	$\dfrac{hp \times 33{,}000}{rpm \times 2\pi} = \dfrac{5250\ hp}{rpm}$	USCS
Density	$\dfrac{Mass}{Unit\ volume}$	ρ	$\dfrac{lb}{ft^3} \times \dfrac{s^2}{ft} = \dfrac{slugs}{ft^3}$	SI
	$\dfrac{Mass}{Unit\ volume}$	ρ	$\dfrac{g}{cm^3} \times \dfrac{s^2}{cm}$	
Absolute viscosity in USCS units	$\dfrac{Mass}{Length \times Time}$	μ	$\dfrac{slugs}{ft \cdot s} = \dfrac{lb \cdot s}{ft^2}$	USCS
			1 unit of $\dfrac{slugs}{ft \cdot s} = 178.69\ P$	
		μ_A	$\dfrac{lb \cdot min}{in^2} = 4{,}136{,}000\ P$	
			$\dfrac{poundal \cdot s}{ft^2} = 14.88\ P$	

Property	Unit/Dimension	Symbol	Formula / Value	System
Absolute viscosity in SI units	P		$\dfrac{1 \text{ dyn} \cdot \text{s}}{\text{cm}^2}$	SI
	cP	Z	$\dfrac{1}{100}$ P	
	$\dfrac{g \cdot s}{cm^2}$		981 P	
Kinematic viscosity	$\dfrac{\text{Area}}{\text{Time}}$	ν	$\nu = \dfrac{\mu}{\rho} = \dfrac{\text{absolute viscosity}}{\text{density}}$	USCS and SI
	St, $\dfrac{cm^2}{s}$		$\dfrac{1 \text{ P}}{\text{density}}$	SI
	cSt		$\dfrac{1}{100}$ St	
	Saybolt universal seconds	V	For conversion of SUS units into centistokes	SI
			When SUS $\leqq$ 100 cSt = 0.226 SUS. − 195/SUS.	
			When SUS. = 100 cSt = 0.220 SUS. − 135/SUS.	
Specific viscosity	Dimensionless		Ratio of absolute viscosity of any fluid to that of water at a temperature of 20° C	Absolute value
Viscosity of water at a temperature of 20°C	cP	Z	Z = 1 cP	SI

TABLE 5.4 Hydrodynamic Formulas Useful in Bearing Design (*Continued*)

Name	Unit	Symbol	Formula or value	System
Fluidity	$\dfrac{\text{Length} \times \text{time}}{\text{Mass}}$	ϕ	$\phi = \dfrac{1}{\mu} =$ inverse value of absolute viscosity	SI and USCS
				Absolute value
Reynold's number	Dimensionless	N_R	$N_R = \dfrac{\rho v d}{\mu} = \dfrac{vd}{\nu}$	
			v = velocity of fluid	
			ρ = density	
			d = pipe diameter	
			μ = absolute viscosity	
			ν = kinematic viscosity	
Critical Reynold's number for pipe flow	Dimensionless	N_e	$N_e = 2320 = $ Reynold's number which represents the separation point between laminar and turbulent flow	Absolute value
			For $N_R > 2320$, turbulent flow	
			For $N_R < 2320$ laminar flow	
			From $N_R = 1920$ to 4000, instability flow	
Friction loss formula for pipe flow	ft or m	H_f	$H_f = f \times \dfrac{l}{d} \times \dfrac{v^2}{2g}$	SI and USCS
			v = velocity, ft/s or m/s	
			d = pipe dia., ft or m	
			l = pipe length, ft or m	
			g = gravity constant, ft/s^2 or m/s^2	
			f = flow coefficient	
Flow coefficient for laminar (viscous) flow	Dimensionless	f	$f = \dfrac{64}{N_R}$	Absolute value
			Grade or roughness of pipe is immaterial	

Flow coefficient for clean cast-iron pipe—circular section	f	$f = \dfrac{0.214}{\sqrt[4]{N_R}}$ turbulent flow $f = \dfrac{64}{N_R}$ laminar flow	Dimensionless	Absolute value
Flow coefficient for very smooth pipe, circular section	f	$f = \dfrac{0.316}{\sqrt[4]{N_R}}$ turbulent flow $f = \dfrac{64}{N_R}$ laminar flow	Dimensionless	Absolute value
Flow coefficient for maximum degree of roughness	f	$f = \dfrac{0.316}{\sqrt[4]{N_R}}$ turbulent flow $\quad = 0.054$ $f = \dfrac{64}{N_R}$ laminar flow	Dimensionless	Absolute value
Reynold's number for water of 20°C temperature	N_R	$N_R = 100vd$ v = velocity, cm/s d = pipe dia., cm	Dimensionless	Absolute value

Bearing constant (Sommerfeld variable)	S	$S = \left(\dfrac{D}{c}\right)^2 \dfrac{\mu n}{p}$
		
		D = bearing diameter
		c = bearing clearance
		   = bearing diameter
		    − shaft diameter
		μ = absolute viscosity
		p = unit pressure
		n = revolutions per time unit
			Dimensionless	Absolute value

TABLE 5.4 Hydrodynamic Formulas Useful in Bearing Design (*Continued*)

Name	Unit	Symbol	Formula or value	System
Example Find Sommerfeld variable for $D/$ $C = 1000$	Dimensionless	S	$S = 1,000,000 \times \dfrac{\mu n}{p}$ $\quad = \dfrac{1,000,000 Zn}{4,136,000 \times 100 \times p}$ μ_A = absolute viscosity, $\qquad$ lb $\cdot$ min/in^2 n = rpm p = lb/in^2 Z = absolute viscosity, cSt For $\dfrac{Zn}{p} = 36,$ $S = 0.00242 \times 36 = 0.087$ *Note:* Burwell's charts show maximum allowable pressure at $s = 0.267$ but reasonably high pressure and minimum coefficient of friction at range $\qquad S = 0.080$ to $S = 0.50$ (See ASME *Transactions* 1942, p. 457.)	Absolute value
Bearing constant	Dimensionless	S_z	$S_z = \dfrac{Zn}{p}$ Z = absolute viscosity, cP n = rpm p = unit pressure, lb/in^2	Absolute value

SECTION 6
METALWORKING FORMULAS

CUTTING SPEED

The cutting speed is important in all machine tool computations. The formula for finding the cutting speed when the rotation speed, denoted by rpm, and the diameter are known is

$$CS = \frac{(D\pi)(\text{rpm})}{12} \qquad (6.1)$$

where CS = cutting speed, ft/min (m/min)
    D = diameter, in (mm)
    π = 3.1416
   rpm = rotation speed, rpm

For all practical purposes, a simple formula can be employed that will give us a "good enough" answer. This is

$$CS = \frac{(D)(\text{rpm})}{4} \qquad (6.2)$$

To find the rpm required when the cutting speed and diameter are known, use

$$\text{rpm} = \frac{4CS}{D} \qquad (6.3)$$

TOOL FEED

The formula for determining the feed of a turning tool, drill, boring tool, reamer, etc., when the length of the cut, the time to make the cut, and the rpm are known, is

$$F = \frac{L}{(\text{rpm})(T)} \tag{6.4}$$

where F = feed, in/rev (mm/rev)
 L = length of cut, in (mm)
 T = time, min

SHAFT TAPERS

To determine the length of a shaft taper or the taper per unit length of shaft, use

$$L = \frac{12(D_1 - D_2)}{T} \tag{6.5}$$

where L = length of taper
 D_1 = larger diameter
 D_2 = smaller diameter
 T = taper per foot (meter)

The taper dimensions are shown in Fig. 6.1.

MILLING MACHINES

In a milling machine the table feed must be adjusted to meet the cutter rpm and feed per tooth per revolution of the cutter. The formula for table feed is

Table feed = (feed per tooth)(number of teeth)(rpm) $\tag{6.6}$

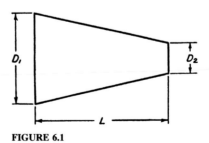

FIGURE 6.1

To figure the total travel of a mill table in making mill cuts, it is necesary to know the length of the approach or the overrun. The formula for calculating the approach on facing cuts is

$$\text{Approach} = \frac{D - \sqrt{D^2 - F^2}}{2} \qquad (6.7)$$

where approach = approach, in (mm)
D = diameter of cutter
F = face of cut

The feed per tooth for a milling cutter is given by

$$\text{Feed per tooth} = \frac{\text{table feed}}{(\text{rpm})(\text{no. of teeth})} \qquad (6.8)$$

Additional machine time must be allowed in face milling. When a milling cutter has traveled the length of the face, some portion of the surface has yet to be milled. This is shown in the shaded area of Fig. 6.2. To complete the millng of the face, an additional distance must be traveled by the milling table. This additional distance must be calculated and a time allowance made for it. The formula for calculating this added table travel, or cutter overrun, is

$$\text{Added table travel} = \tfrac{1}{2}(D - \sqrt{D^2 - W^2}) \qquad (6.9)$$

For milling machine calculations there are three formulas involved. The first is

$$\text{CS} = \frac{(\text{rpm})(D)}{4} \qquad (6.10)$$

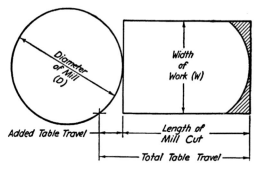

FIGURE 6.2 Cutter approach or cutter overrun.

140 SECTION SIX

where CS = cutting speed, ft/min (m/min)
rpm = rotation speed, rpm
D = diameter of cutter, in (mm)

The second formula is

$$rpm = \frac{4CS}{D} \tag{6.11}$$

The third is

$$Time = \frac{length\ of\ cut}{(feed\ per\ tooth)(no.\ of\ teeth)(rpm)} \tag{6.12}$$

MILLING CUTTER OVERRUN

In most milling operations the table travel exceeds the length of the milled dimension shown on the part drawing. The formula for the overrun (Fig. 6.3) is

$$Overrun = \sqrt{Dd - d^2} \tag{6.13}$$

where D = diameter of cutter, in (mm), and d = depth of cut, in (mm). This formula holds true only where the depth of the cut is less than the radius of the cutter. When the depth of the cut equals or exceeds the radius of the cutter, the overrun will always equal the radius of the cutter.

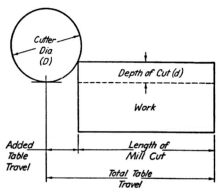

FIGURE 6.3 Cutter overrun.

DEPTH OF MILLING CUT REQUIRED FOR A KEY SEAT

The depth of a milling cut for a key seat is given by

$$D = \frac{T}{2} + A \qquad (6.14)$$

$$A = R - \sqrt{R^2 - \left(\frac{w}{2}\right)^2} \qquad (6.15)$$

where D = dept of cut in (mm)
   T = thickness of key
   W = width of key
   A = height of arc
   R = radius of shaft

where the dimensions are as shown in Fig. 6.4.

PRODUCTION TIME

The formula for calculating the true unit production time for various lot sizes is

$$T = \frac{SU}{N} + U \qquad (6.16)$$

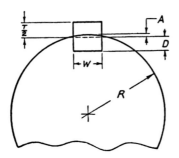

FIGURE 6.4   Key and key seat dimensions.

where T = true unit time, min
  SU = setup time, min
  N = number of pieces in lot
  U = unit standard time, min

BEST PRODUCTION METHOD

When more than one method to produce a part is available, the number of pieces of the part for which a change of method is desirable can be found from

$$N = \frac{SU_2 - SU_1}{T_1 - T_2} \tag{6.17}$$

where N = number of pieces for which a change of method is desirable
  SU_1 = setup time of first method
  SU_2 = setup time of second method
  T_1 = unit time by first method
  T_2 = unit time by second method

MINIMUM LOT SIZE

The formula for determining the minimum lot size based upon the setup and the unit standard is

$$N = \frac{SU}{TK} \tag{6.18}$$

where N = minimum lot size
  SU = setup time
  T = unit standard time
  K = arbitrarily selected maximum percent

TIME TO TURN

There are a number of factors that influence the time it takes to turn stock from one diameter to another. The diameter of the part, the kind of material, the spindle speed, the feed and the depth of cut all influence time. If the spindle speed (or the cutting speed) is one-half of what it should be, the time to make the cut will be twice as long. The same holds true with the feed. If the feed and speed both are one-half of what they should be, the time to make the cut will be 4 times as long.

$$\text{Time} = \frac{\text{length of cut}}{(\text{feed})(\text{rpm})} \tag{6.19}$$

TIME REQUIRED TO CHANGE TOOLS

The formula for unit time to change tools is

$$\text{UT} = \frac{(\text{TT})(\text{CT})}{L} \tag{6.20}$$

where UT = unit time to change from dull to sharp tools
TT = total time to change tool
CT = cutting time tool is in use during operation cycle
L = life of tool

CHAMFERING TIME

The formula for calculating the time it takes to chamfer is

$$\text{Time to chamfer} = \frac{\text{length of cut}}{(\text{feed})(\text{rpm})} \tag{6.21}$$

TIME TO FACE

The time to face a part on a lathe is given by

$$\text{Time to face} = \frac{\text{length of cut}}{(\text{feed})(\text{rpm})} \tag{6.22}$$

THREAD DIMENSIONS NEEDED WHEN CUTTING THREADS

The depth of a thread is an important element in estimating thread cutting time, horsepower (kW), production rate, and other key variables. The depth of various screw threads is calculated as follows:

1. American standard V thread

$$d = \frac{0.6495}{N} \tag{6.23}$$

where d = depth of thread, in (m), and N = number of threads per inch (threads per millimeter).

2. Acme thread

$$d = \frac{0.500}{N} + 0.010 \quad \text{in} \tag{6.24}$$

3. Square thread (Fig. 6.5)

$$d = \frac{0.500}{N} \tag{6.25}$$

4. A 20° worm, thread (Fig. 6.5)

$$d = \frac{0.6866}{N} \tag{6.26}$$

5. Buttress thread

$$d = \frac{0.750}{N} \tag{6.27}$$

The whole depth of a gear—usually representing the total depth of the tooth—is calculated from the following formulas:

1. Standard spur gear

$$D = \frac{2.157}{P} \tag{6.28}$$

where D = whole depth of tooth and P = diametral pitch.

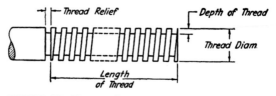

FIGURE 6.5   Elements of acme square or worm thread.

2. Bevel gear

$$D = \frac{2.157}{P} \qquad (6.29)$$

3. Spiral bevel gear

$$D = \frac{1.888}{P} \qquad (6.30)$$

4. Worm gear

$$D = 0.6866P \qquad (6.31)$$

where P = linear pitch.

5. Spiral gear

$$D = \frac{2.157}{P} \qquad (6.32)$$

where P = normal diametral pitch.

THREADING OPERATIONS

The formula for time to thread is

$$\text{Time} = \frac{(\text{length of cut})(\text{threads per inch})}{\text{rpm}} \qquad (6.33)$$

(In the SI thread system, it is threads per millimeter.)

The machine time to cut an acme, a square, or a worm thread can be calculated from the formula

$$T = \frac{LDdn}{4(\text{fpm})(f)} \qquad (6.34)$$

where T = time to machine thread, min
L = length of thread, in (mm)
D = major diameter of thread
d = depth of thread
n = number of threads per inch (millimeter)
fpm = cutting speed, ft/min
f = depth of cut per pass

TIME TO TAP

The time to tap threads in a piece is given by

$$T = \frac{ND\pi}{8(\text{fpm})} \qquad (6.35)$$

where T = time, min, to tap 1.0 in (25.4 mm) deep and reverse-tap for the same distance
     N = number of threads per inch (millimeter)
     D = diameter of the tap, in (mm)
     π = 3.1416
   fpm = cutting speed, ft/min (m/min)

TIME TO PROFILE

The time to profile or route a part is given by

$$\text{Time} = \frac{\text{length of cut}}{(\text{feed per tooth})(\text{no. of teeth})(\text{rpm})} \qquad (6.36)$$

Length of cut and feed can be in either inches or millimeters.

TIME TO CHANGE SHAPER CUTTER

The unit time to change a shaper cutter is given by

$$\text{Unit time} = \frac{(\text{time to change cutter})(\text{cutting time per part})}{\text{life of tool}} \qquad (6.37)$$

SHAPER CUTTING SPEED

There are two helpful formulas for calculating the cutting speed when the number of strokes per minute and the length of the stroke are given or for calculating the number of strokes per minute when the cutting speed and the length of the stroke are known. These are

$$CS = \frac{NL}{6} \qquad (6.38)$$

$$N = \frac{6(CS)}{L} \qquad (6.39)$$

where N = number of strokes per minute of ram
L = length of stroke
CS = cutting speed, ft/min (m/min)

Shaper Cutting Time

The time to cut a piece on a shaper (Fig. 6.6) is given by

$$\text{Time} = \frac{\text{length of cut}}{(\text{feed})(\text{strokes per minute})} \qquad (6.40)$$

This formula can also be used to compute the time to plane.

FEED RATE FOR CENTER-TYPE GRINDER

The formula for the feed rate of the work is

$$F = \pi dN \sin \alpha \qquad (6.41)$$

where F = feed of work, in/min (mm/min)
d = diameter of regulating wheel, in (mm)
N = speed of regulating wheel, rpm
α = angle of inclination of regulating wheel

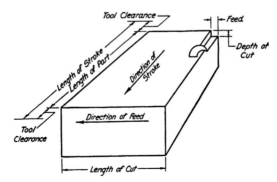

FIGURE 6.6 Shaper operation terms.

GRINDING TIME

The following formulas will give approximate grinding times for the average work done in a machine shop.

For roughing cuts,

$$T = \frac{LT_s D}{(w/2)F \; (4CS)} \quad \text{or} \quad T = \frac{LT_s D}{2w \, F(CS)} \tag{6.42}$$

For finishing cuts,

$$T = \frac{LT_s D}{w \, F(CS)} \tag{6.43}$$

where T = time, min
L = length of part
T_s = total amount of stock
D = diameter of part
w = width of face of grinding wheel
F = feed or depth of cut
CS = cutting speed, ft/min (m/min)

The constant $\frac{1}{2}$ in the roughing formula represents a longitudinal feed of one-half the wheel face per revolution of the work. And T_s/F represents the number of cuts necessary to remove the total stock. And $D/4CS$ is the reciprocal of $4CS/D$ in the rpm formula.

TIME TO THREAD MILL

The time to thread mill is given by

$$\text{Time} = \frac{\text{length of cut}}{(\text{feed per flute})(\text{no. of flutes})(\text{rpm})} \tag{6.44}$$

TIME REQUIRED TO BROACH

In general, the formula for time to broach is

$$\text{Time} = \frac{\text{length of stroke}}{\text{cutting speed}} + \frac{\text{length of stroke}}{\text{return speed}} \tag{6.45}$$

HOB SPEED AND WORK SPEED

The formulas for hob speed and work speed are

$$HR = (WR)(N) \qquad (6.46)$$

or
$$WR = \frac{HR}{N} \qquad (6.47)$$

where WR = rotation speed of work spindle, rpm
  HR = rotation speed of hob, rpm
   N = number of gear teeth, splines, or serrations

TIME TO MACHINE SPUR GEARS

The formula for estimating the time to machine a spur gear is

$$T = \frac{N(L + A)}{F(\text{rpm})} \qquad (6.48)$$

where T = cutting time, min
   N = number of teeth in gear to be cut
   L = face of gear or length of tooth
   A = approach of hob—distance hob travels before reaching full
       depth of cut
   F = feed of hob, in/r (mm/r) of hob
  rpm = rotation speed of hob, rpm

The number of teeth to be cut and the length of the gear face are known factors that can be gathered from the blueprint. The approach of the cutter can be calculated. The feed of the hob and the rpm of the hob will be found in tables of recommendations.
The formula for calculating the approach of the hob to the work is

$$A = \sqrt{d(D - d)} \qquad (6.49)$$

where A = approach, in (mm)
   D = diameter of hob, in (mm)
   d = whole depth of gear tooth, in (mm)

HOBBING TIME FOR SHAFTS

The time to machine serrations on a shaft by the hobbing method is calculated by

$$T = \frac{N(L + A)}{KF(\text{rpm})} \tag{6.50}$$

where T = time, min
  N = number of serrations
  L = length of serrations
  A = approach, in (mm)
  K = feed per flute, in (mm)
  F = number of flutes on hob
  rpm = rotation speed of hob, rpm

TIME TO SAW FOR POWER HACKSAW

The formula is

$$\text{Time to saw} = \frac{\text{length of cut}}{(\text{strokes per minute})(\text{feed per stroke})} \tag{6.51}$$

TIME FOR BAND SAW CUTTING

$$\text{Time to saw} = \frac{\text{length of cut, in (mm)}}{[\text{cutting speed, ft/min (m/min)}](\text{pitch of saw})(\text{feed/tooth})} \tag{6.52}$$

The pitch of the saw = teeth per inch (millimeter).

HORSEPOWER REQUIRED AT MOTOR TO MAKE TURNING CUT(S)

Carboloy Company gives this formula:

	Motor hp
Machine A	Cut B
Power required to overcome friction in the machine, usually figured as 30% of B	Calculate horsepower required by each tool, using formula given below:

	Tools in cut at one time hp
	  Tool 1 (DFSC) =
	  Tool 2 (DFSC) =
	  Etc. =

	Total horsepower required by tools in cut B =

TABLE 6.1 Power Constants for Various Metals

Material	Power constant	Material	Power constant
Magnesium alloy	3	Cast iron, medium	4
Aluminum castings	3	Cast iron, soft	3
Aluminum bar stock	4	Copper	4
Brass, hard	10	SAE X1112	3
Brass, soft	4	SAE X4130	4
Bronze, hard	10	Stainless steel	4
Bronze, soft	4	Monel metal	5
Cast iron, hard	5		

The following formula is used for the above calculations:

$$\text{hp} = DFSC \tag{6.53}$$

where D = depth of cut, in (mm)
F = feed, in (mm)
S = surface speed, ft/min (m/min)
C = power constant, Table 6.1

Note: hp $\times$ 0.746 = kW.

SECTION 7
FORMULAS FOR HEATING, VENTILATING, AND AIR CONDITIONING

AIR CONDITIONING FORMULAS

Fan Laws

For a centrifugal fan, the airflow delivered in cubic feet per minute (CFM) (m^3/min), varies with the rotational speed RPM of the fan wheel, or

$$\frac{CFM_2}{CFM_1} = \frac{RPM_2}{RPM_1} \tag{7.1}$$

The static pressure SP, in (mm) of water, varies with the airflow delivered (CFM) or RPM thus:

$$\frac{SP_2}{SP_1} = \left(\frac{CFM_2}{CFM_1}\right)^2 = \left(\frac{RPM_2}{RPM_1}\right)^2 \tag{7.2}$$

Power input to the fan, expressed in brake horsepower bhp (kW), to deliver the required airflow, varies with the airflow, RPM, and static pressure thus:

$$\frac{BHP_2}{BHP_1} = \left(\frac{CFM_2}{CFM_1}\right)^3 = \left(\frac{RPM_2}{RPM_1}\right)^3 = \left(\frac{SP_2}{SP_1}\right)^{1.5} \tag{7.3}$$

The brake horsepower input needed to deliver a given airflow with a fan efficiency FAN_{eff} of 65 to 85 percent, expressed as a decimal, is

$$BHP = \frac{CFM \times SP \times SP.GR.}{6356 \times FAN_{eff}} \tag{7.4}$$

With an electric motor-driven M/D fan (the most common drive used today) having an efficiency M/D_{eff} of 80 to 95 percent,

$$MHP = \frac{BHP}{M/D_{eff}}$$ (7.5)

In these formulas the fan size, i.e., wheel diameter, is assumed constant, as is the air density during fan operation. The specific gravity SP.GR. of the air being moved $= 1.0$.

Air Change Formulas

The air change rate per hour, AC/HR, with an airflow CFM, ft^3/min (m^3/min), and a room or space VOLUME, ft^3 (m^3) is

$$\frac{AC}{HR} = \frac{CFM \times 60}{VOLUME}$$ (7.6)

Or, with a known air change per hour for a space having a stated volume, the airflow required is

$$CFM = \frac{AC/HR \times VOLUME}{60}$$ (7.7)

Airflow Formulas

The total pressure TP in an air duct, in (mm) of water, with a static pressure SP, in (mm) of water, and a velocity pressure VP, in (mm) of water is

$$TP = SP + VP$$ (7.8)

In a duct the velocity pressure VP, in (mm) of water, with a velocity V of the air, ft/min, is

$$VP = \left(\frac{V}{4005}\right)^2 = \frac{V^2}{4005^2}$$ (7.9)

Note: in of water $\times$ 2.54 = mm of water.

The air velocity V, ft/min, in an air duct with a flow rate of Q ft^3/min through a duct having a cross-sectional area of A ft^2 is

$$V = \frac{Q}{A} = \frac{Q \times 144}{W \times H}$$ (7.10)

Note: ft/min $\times$ 0.3408 = m/min.

To find the equivalent diameter D_{EQ} of a round duct for a known rectangular duct size having a width of A in and a height of B in, use

$$D_{EQ} = \frac{1.3(AB)^{0.625}}{(A + B)^{0.25}}$$ (7.11)

Formulas for Air Cooling and Heating

The sensible heat H_S in an airstream with a flow rate CFM and a temperature difference of ΔT, °F, is

$$H_S = 1.08 \times \text{CFM} \times \Delta T \tag{7.12}$$

In this same airstream the latent heat H_L with a humidity ratio difference, ΔW_{GR}, gr of dry air/lb water, is

$$H_L = 0.68 \times \text{CFM} \times \Delta W_{GR} \tag{7.13}$$

When the humidity ratio difference is expressed in pounds of water per pound of dry air, the latent heat is

$$H_L = 4840 \times \text{CFM} \times \Delta W_{LB} \tag{7.14}$$

The total heat H_T with an enthalpy difference Δh, Btu/lb of dry air, is

$$H_T = 4.5 \times \text{CFM} \times \Delta h \tag{7.15}$$

Total heat can also be found by taking the sum of sensible and latent heat.

$$H_T = H_S + H_L \tag{7.16}$$

The sensible heat ratio SHR is

$$\text{SHR} = \frac{H_S}{H_T} = \frac{H_S}{H_S + H_L} \tag{7.17}$$

When outside supply air CFM_{OA} is mixed with room return air CFM_{RA} and supply air CFM_{SA}, the resulting mixed air temperature T_{MA} is given by

$$T_{MA} = \left(T_{ROOM} \times \frac{\text{CFM}_{RA}}{\text{CFM}_{SA}}\right) + \left(T_{OA} \times \frac{\text{CFM}_{OA}}{\text{CFM}_{SA}}\right) \tag{7.18}$$

where T_{ROOM} = design room temperature and T_{OA} = outside air temperature. And the mixed air temperature T_{MA} is

$$T_{MA} = \left(T_{RA} \times \frac{\text{CFM}_{RA}}{\text{CFM}_{SA}}\right) + \left(T_{OA} \times \frac{\text{CFM}_{OA}}{\text{CFM}_{SA}}\right) \tag{7.19}$$

Condensate Formed in Air Conditioning

The rate of condensate formation in air conditioning, denoted by $\text{GPM}_{AC\,COND}$, with an airflow rate of CFM, ft³/min, a specific volume of air of SV, ft³/lb of dry air, and a specific change of ΔW_{LB}, lb of water/lb of dry air, is

$$\text{GPM}_{\text{AC COND}} = \frac{\text{CFM} \times \Delta W_{\text{lb}}}{\text{SV} \times 8.33} \qquad (7.20)$$

With specific humidity ΔW_{GR}, gr of water/lb of dry air, is

$$\text{GPM}_{\text{AC COND}} = \frac{\text{CFM} \times \Delta W_{\text{GR}}}{\text{SV} \times 8.33 \times 7000} \qquad (7.21)$$

And the steam flow required, lb/h, is, with H_{FG} the latent heat of vaporization, Btu/lb, at the design pressure of the system

$$\frac{\text{LB STM}}{\text{HR}} = \frac{\text{BTU/HR}}{H_{FG}} \qquad (7.22)$$

Overall heat transfer H, Btu/h, through an area of A ft^2 with a temperature difference of ΔT, °F, is

$$H = U \times A \times \Delta T \qquad (7.23)$$

Formulas for Balancing Air Needs

With airflows of cubic feet (meters) per minute, the supply air SA, in terms of return air RA, outside air OA, exhaust air EA, and relief air RFA, is then

$$\text{SA} = \text{RA} + \text{OA} = \text{RA} + \text{EA} + \text{RFA} \qquad (7.24)$$

When the minimum outside ventilation air OA exceeds the exhaust air EA, the outside air is

$$\text{OA} = \text{EA} + \text{RFA} \qquad (7.25)$$

Where the exhaust air EA exceeds the minimum outside ventilation air OA, the outside air is

$$\text{OA} = \text{EA} \qquad \text{RFA} = 0 \qquad (7.26)$$

With an economizer cycle in the design,

$$\text{OA} = \text{SA} = \text{EA} + \text{RFA} \qquad \text{RA} = 0 \qquad (7.27)$$

Formulas for Room Humidification

The grains of moisture required to deliver the required humidification in a room is the difference between grains of water per cubic foot of room air, with a specific humidity of W grains per pound of dry air and a specific volume of air SV, ft^3/lb of dry air, and the same for the supply air. Or

$$\text{GRAINS}_{\text{REQ'D}} = \left(\frac{W_{\text{GR}}}{\text{SV}}\right)_{\text{ROOM AIR}} - \left(\frac{W_{\text{GR}}}{\text{SV}}\right)_{\text{SUPPLY AIR}} \qquad (7.28)$$

And the pounds of moisture required for humidification, when the specific humidity is expressed in pounds of water per pound of dry air, is

$$\text{POUNDS}_{\text{REQ'D}} = \left(\frac{W_{\text{LB}}}{\text{SV}}\right)_{\text{ROOM AIR}} - \left(\frac{W_{\text{LB}}}{\text{SV}}\right)_{\text{SUPPLY AIR}} \qquad (7.29)$$

The pounds of steam required per hour to produce the desired humidity is

$$\text{LB STM/HR} = \frac{\text{CFM} \times \text{GRAINS}_{\text{REQ'D}} \times 60}{7000}$$

$$= \text{CFM} \times \text{POUNDS}_{\text{REQ'D}} \times 60 \qquad (7.30)$$

To determine the humidifier sensible heat gain H_S, Btu/h, with steam flow of Q lb/h and a temperature difference of T between the steam and the supply air, and a humidifier manifold length of L ft, use

$$H_S = 0.244 \times Q \times \Delta T + L \times 380 \qquad (7.31)$$

Determining the Condensation Temperature of Moisture on Glass Windows

Moisture in the air will condense on glass windows when the temperature of the glass T_{GLASS} is less than the dew point DP_{ROOM} temperature of the room air. Or,

$$T_{\text{GLASS}} = T_{\text{ROOM}} - \left[\frac{R_{\text{IA}}}{R_{\text{GLASS}}} \times (T_{\text{ROOM}} - T_{\text{OA}})\right] \qquad (7.32)$$

where T = temperature, °F
  R = R value, h·ft²·°F/Btu
  U = U value, Btu/(h·ft²·°F)
  IA = inside air film
  OA = design outside air temperature
  DP = dew point

When you use the overall U value for the glass,

$$T_{\text{GLASS}} = T_{\text{ROOM}} - \left[\frac{U_{\text{GLASS}}}{U_{\text{IA}}} \times (T_{\text{ROOM}} - T_{\text{OA}})\right] \qquad (7.33)$$

If $T_{\text{GLASS}} < \text{DP}_{\text{ROOM}}$, condensation occurs.

Formulas for the Properties of Air in Air Conditioning

Specific humidity W of air used in air conditioning, lb (gr) of water/lb of dry air, is given by

$$W = 0.622 \times \frac{P_W}{P - P_W} \tag{7.34}$$

$$W = \frac{(2501 - 2.381T_{WB})(W_{SATWB}) - (T_{DS} - T_{WB})}{2501 + 1.805T_{DB} - 4.186T_{WB}} \tag{7.35a}$$

$$W = \frac{(1093 - 0.556T_{WB})(W_{SATWB}) - 0.240(T_{DB} - T_{WB})}{1093 + 0.444T_{DB} - T_{WB}} \tag{7.35b}$$

where W = specific humidity, lb H_2O/lb DA or gr H_2O/lb DA
W_{ACTUAL} = actual specific humidity, lb H_2O/lb DA or gr H_2O/lb DA
W_{SAT} = saturation specific humidity at dry-bulb temperature
W_{SATWB} = saturation specific humidity at wet-bulb temperature
P_W = partial pressure of water vapor, lb/ft^2
P = total absolute pressure of air/water vapor mixture, lb/ft^2
P_{SAT} = saturation partial pressure of water vapor at dry-bulb temperature, lb/ft^2
RH = relative humidity, %
H_S = sensible heat, Btu/h
H_L = latent heat, Btu/h
H_T = total heat, Btu/h
m = mass flow rate, lb/h DA or H_2O
c_P = specific heat (air: 0.24 Btu/lb DA, water: 1.0 Btu/lb H_2O)
T_{DB} = dry-bulb temperature, °F
T_{WB} = wet-bulb temperature, °F
ΔT = temperature difference, °F
ΔW = specific humidity difference, lb H_2O/lb DA or gr H_2O/lb DA
Δh = enthalpy difference, Btu/lb DA
L_v = latent heat of vaporization, Btu/lb H_2O

Relative humidity RH of the air is

$$RH = \frac{W_{ACTUAL}}{W_{SAT}} \times 100\% \tag{7.36}$$

$$RH = \frac{P_W}{P_{SAT}} \times 100\% \tag{7.37}$$

The sensible, latent, and total heats are

$$H_S = mc_P \times \Delta T \tag{7.38}$$

$$H_L = L_v m \times \Delta W \tag{7.39}$$

$$H_T = m \times \Delta h \tag{7.40}$$

Chilled-Water System Formulas

The total heat H removed by a chilled-water installation in an air conditioning system is given by

$$H = 500 \times \text{GPM} \times \Delta T \tag{7.41}$$

The evaporator water flow rate GPM_{EVAP} is

$$\text{GPM}_{\text{EVAP}} = \frac{\text{TONS} \times 24}{\Delta T} \tag{7.42}$$

Condenser flow rate GPM_{COND} is

$$\text{GPM}_{\text{COND}} = \frac{\text{TONS} \times 30}{\Delta T} \tag{7.43}$$

where H = total heat, Btu/h
 GPM = water flow rate, gal/min
 ΔT = temperature difference, °F
 TONS = air conditioning load, tons
 GPM_{EVAP} = evaporator water flow rate, gal/min
 GPM_{COND} = condenser water flow rate, gal/min

Refrigeration System Cooling Tower Formulas

For cooling towers and heat exchangers the approach temperature $\text{APPROACH}_{\text{CT'S}}$ for the cooling tower and $\text{APPROACH}_{\text{HE'S}}$ for the heat exchanger are important in evaluating performance. These temperatures are given by

$$\text{APPROACH}_{\text{CT'S}} = \text{LWT} - \text{AWB} \tag{7.44}$$

$$\text{APPROACH}_{\text{HE'S}} = \text{EWT}_{\text{HS}} - \text{LWT}_{\text{CS}} \tag{7.45}$$

$$\text{RANGE} = \text{EWT} - \text{LWT} \tag{7.46}$$

where EWT = entering water temperature (°F)
 LWT = leaving water temperature (°F)
 AWB = ambient wet-bulb temperature (design WB, °F)
 HS = hot side
 CS = cold side

For the cooling tower itself,

$$C = \frac{E + D + B}{D + B} \tag{7.47}$$

$$B = \frac{E - (C - 1)D}{C - 1} \tag{7.48}$$

$$E = \text{GPM}_{\text{COND}} \times R \times 0.0008 \tag{7.49}$$

$$D = \text{GPM}_{\text{COND}} \times 0.0002 \tag{7.50}$$

$$R = \text{EWT} - \text{LWT} \tag{7.51}$$

where B = blowdown, gal/min
C = cycles of concentration
D = drift, gal/min
E = evaporation, gal/min
EWT = entering water temperature, °F
LWT = leaving water temperature, °F
R = range, °F

HEATING SYSTEM FORMULAS

Pressure Loss in Steam Piping

Consider a condition of steady flow in a pipe. Let p_1 (Fig. 7.1) be the unit static pressure of the fluid at one point and let p_2 be the pressure at another point at a distance L from the first. The drop in pressure due to the friction of the fluid in passing through the distance L is then

$$P = p_1 - p_2 \tag{7.52}$$

Expressing the laws of friction stated above in algebraic form, we have

$$Pa = FSDv^2 \tag{7.53}$$

where P = drop in unit pressure, lb/ft²
a = cross-sectional area of pipe, ft²
F = constant depending on nature of fluid and nature of pipe surface
S = area of contact between fluid and pipe, ft²
D = density of the fluid, lb/ft³
v = velocity of the flow, ft/s

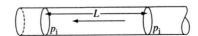

FIGURE 7.1

Then

$$P = \frac{1}{a} FSDv^2 \tag{7.54}$$

Let F be made arbitrarily $= f/2g$. Then Eq. (7.54) becomes

$$P = \frac{1}{a} fSD \frac{v^2}{2g} \tag{7.55}$$

For round pipes of diameter d and length L, $S = \pi dL$, and $a = \pi d^2/4$. Then

$$P = \frac{4fLDv^2}{d^2 g} \tag{7.56}$$

Let w = the weight of steam flowing in pounds per minute. Then

$$w = \frac{\pi d^2}{4} \times vD \times 60 = 47.12 d^2 vD \tag{7.57}$$

and

$$v = \frac{w}{47.12 d^2 D} \tag{7.58}$$

Let p be the pressure drop in pounds per square inch $= P/144$, and let d_1 be the diameter in inches $= 12d$. Substituting gives

$$p = 0.04839 \frac{fw^2 L}{Dd_1^5} \tag{7.59}$$

The coefficient f was found by Unwin to be

$$f = K\left(1 + \frac{3}{10d}\right) = K\left(1 + \frac{3.6}{d_1}\right) \tag{7.60}$$

The value most commonly used for K for steam is that determined by Babcock 0.0027. Substituting gives

$$p = 0.0001306 w^2 L \frac{1 + 3.6/d_1}{Dd_1^5} \tag{7.61}$$

where p = pressure drop, lb/in^2
$\quad\quad w$ = weight of steam flowing, lb/min
$\quad\quad L$ = length of pipe, ft
$\quad\quad d_1$ = diameter of pipe, in
$\quad\quad D$ = *average* density of steam, lb/ft^3

Formulas for Sizing Domestic Hot-Water Heaters

$$H_{\text{OUTPUT}} = \text{GPH} \times 8.34 \text{ LBS/GAL} \times \Delta T \times 1.0 \quad (7.62)$$

$$H_{\text{INPUT}} = \frac{\text{GPH} \times 8.34 \text{ LBS/GAL} \times \Delta T}{\% \text{ EFFICIENCY}} \quad (7.63)$$

$$\text{GPH} = \frac{H_{\text{INPUT}} \times \% \text{ EFFICIENCY}}{\Delta T \times 8.34 \text{ LBS/GAL}}$$

$$= \frac{\text{KW} \times 3413 \text{ BTU/KW}}{\Delta T \times 8.34 \text{ LBS/GAL}} \quad (7.64)$$

$$\Delta T = \frac{H_{\text{INPUT}} \times \% \text{ EFFICIENCY}}{\text{GPH} \times 8.34 \text{ LBS/GAL}}$$

$$= \frac{\text{KW} \times 3413 \text{ BTU/KW}}{\text{GPH} \times 8.34 \text{ LBS/GAL}} \quad (7.65)$$

$$\text{KW} = \frac{\text{GPH} \times 8.34 \text{ LBS/GAL} \times \Delta T \times 1.0}{3413 \text{ BTU/KW}} \quad (7.66)$$

$$\% \text{ COLD WATER} = \frac{T_{\text{HOT}} - T_{\text{MIX}}}{T_{\text{HOT}} - T_{\text{COLD}}} \quad (7.67)$$

$$\% \text{ HOT WATER} = \frac{T_{\text{MIX}} - T_{\text{COLD}}}{T_{\text{HOT}} - T_{\text{COLD}}} \quad (7.68)$$

where H_{OUTPUT} = heating capacity, output
$ H_{\text{INPUT}}$ = heating capacity, input
$ \text{GPH}$ = recovery rate, gal/h
$ \Delta T$ = temperature rise, °F
$ \text{KW}$ = kilowatts
$ T_{\text{COLD}}$ = temperature, cold water, °F
$ T_{\text{HOT}}$ = temperature, hot water, °F
$ T_{\text{MIX}}$ = temperature, mixed water, °F

Heating Capacity of Radiators and Convectors

The heating capacity of a radiator is

Btu emission per hour at 70° room and 215° steam temperatures

$$= \text{lb/h of condensation} \times 970 \times \text{correction factor} \quad (7.69)$$

Correction factor C_s

$$= \left(\frac{215 - 70}{\text{average steam temperature} - \text{room temperature}}\right)^{1.3} \quad (7.70)$$

Heat emission of a steam-heated convector is given by Eq. (7.69) with a correction factor of

$$C_s = \left(\frac{215 - 65}{t_s - t_i}\right)^{1.5} \quad (7.71)$$

where C_s = correction factor, to be applied in Eq. (7.69)
t_s = steam temperature in test
t_i = average inlet air temperature in test

The exponent 1.5 has been shown by test to be the proper figure for convectors. With hot-water convectors the correction factor to use is

$$C_w = \left[\frac{\theta_s - 65}{(\theta_1 + \theta_2)/2 - t_i}\right]^{1.5} \quad (7.72)$$

where C_w = correction factor, as before
θ_s = one of standard mean water temperatures 170, 190, 210, or 230°
θ_1 = inlet water temperature, deg
θ_2 = outlet water temperature, deg

Room Air Supply Based on CO_2 Content

The air supply per occupant is calculated from the CO_2 measurement as follows, assuming that the CO_2 produced per occupant, denoted by CFH, is 0.6 ft^3/h:

$$\frac{\text{CFH}(CO_2 - X)}{10,000} = 0.6 \quad (7.73)$$

where CFH = air supplied to room per occupant, ft^3/h
CO_2 = carbon dioxide content of room air, parts per 10,000
X = carbon dioxide content of outside air, parts per 10,000 (usually assumed as 4)

Then

$$\text{CFH} = \frac{6000}{CO_2 - X} \quad (7.74)$$

Fan Air and Input Horsepower and Efficiency

The power required for moving air through a system of ducts may be expressed as follows. Let

$$p = \text{total pressure, in } H_2O$$

$$a = \text{cross-sectional area of duct, ft}^2$$

$$v = \text{velocity of air, ft/min}$$

Then, applying the necessary conversion factors to change the total pressure to pounds per square foot, we get

$$\text{Ahp} = \frac{pav \times 144}{12 \times 2.31 \times 33,000} \qquad (7.75)$$

or
$$\text{Ahp} = 0.000157 pav \qquad (7.76)$$

where Ahp is air horsepower. If q is the volume of air delivered per minute in cubic feet, then $q = av$, and

$$\text{Ahp} = 0.000157 pq = \frac{pq}{6356} \qquad (7.77)$$

in which the pressure p is expressed in inches of water.

If the pressure is expressed in terms of the equivalent column of air of height, h, then

$$\text{Ahp} = \frac{hDQ}{33,000} \qquad (7.78)$$

in which $D =$ the density of the air, lb/ft^3, and $Q =$ the air volume, ft^3/min.

In a fan the actual head developed is only a portion of the theoretical head v^2/g and is represented approximately by kv^2/g. The input horsepower Hp required to drive a fan is then

$$\text{Hp} = \frac{ckv^2}{g} \times \frac{DQ}{33,000} \qquad (7.79)$$

in which c is a factor which takes into account the mechanical losses in the fan. Combining all the constant factors, we have

$$\text{Hp} = Kv^2QD \qquad (7.80)$$

with v being the peripheral velocity, which varies directly as the speed of the fan. Since Q varies directly as the speed, the power required varies as the cube of the speed.

The fan *static efficiency* is the mechanical efficiency multiplied by the ratio of static pressure to total pressure. The mechanical efficiency ME is expressed by the formula

$$ME = \frac{Ahp}{input\ Hp} \tag{7.81}$$

$$ME = \frac{pq}{6356\ Hp} \tag{7.82}$$

in which Hp = input horsepower.

$$Static\ efficiency\ SE = \frac{pq}{6356\ Hp} \times \frac{static\ pressure}{total\ pressure} \tag{7.83}$$

$$SE = \frac{p_s q}{6356\ Hp} \tag{7.84}$$

in which p_s = static pressure.

Round Duct Diameter Equivalent to a Rectangular Duct

The round duct diameter equivalent in resistance and airflow to a rectangular duct having sides of lengths A and B is

$$D = 1.265 \sqrt[5]{\frac{(AB)^3}{A + B}} \tag{7.85}$$

in which A and B are sides of rectangular duct and D = diameter of round pipe.

Pressure Loss in Air Ducts

The general expression for the friction of fluids in pipes is known as the *Fanning formula* and is approximately applicable to air:

$$P = f \frac{S}{a} D \frac{v^2}{2g} \tag{7.86}$$

where P = pressure required to overcome friction, lb/ft^2
a = cross-sectional area of duct, ft^2
D = density of air, lb/ft^3
v = velocity, ft/s
f = coefficient of friction
S = area of contact (perimeter × length)

For standard air (70° and 29.92-in barometer),

$$P = \frac{0.03FL}{d^{1.24}} \left(\frac{v}{1000} \right)^{1.84} \qquad (7.87)$$

where F = factor for roughness
L = length of duct, ft
d = diameter of duct, in
P and v as above

Air Filter Dust Arrestance

The *dust arrestance E*, usually expressed in percent, is determined by

$$E = 1 - \frac{G_1}{G_0} \qquad (7.88)$$

where E = arrestance
G_1 = dust concentration beyond filter
G_0 = dust concentration ahead of filter

The dust-holding capacity of a nonautomatic filter is the amount by weight of standard dust that the filter will hold without exceeding the following resistances:

0.18 in for low-resistance type

0.50 in for medium-resistance type

1.00 in for high-resistance type

These resistance limits serve to classify filters. The low-resistance type is usually used in warm-air furnace work and in unit air conditioners. The medium-resistance type is usually specified for central fan systems. The high-resistance type is more suitable for such services as the intakes of air compressors than for general air conditioning work.

Heat Gain through a Wall

The heat gain through a wall may be expressed by

$$H = H_t + H_R \qquad (7.89)$$

in which

$$H_t = \text{heat gain due to air temperature difference}$$

$$= AU(t_o - t) \qquad (7.90)$$

where A = area of wall, ft^2
  U = coefficient of heat transmission, Btu/(ft$^2 \cdot$ h $\cdot$ °F)
  t_o = outdoor air temperature, °F
  t = room air temperature, °F

and

$$H_R = AFaI \tag{7.91}$$

where H_R = heat gain due to sun radiation
  A = area of wall or roof, ft^2
  F = portion, expressed as a decimal, of absorbed solar radiation which is transmitted to inside
  a = portion, expressed as a decimal, of impinging solar radiation which is absorbed by wall surface
  I = actual intensity of solar radiation striking the surface, Btu/(h $\cdot$ ft^2)

The values of F are related to the transmission coefficient of the wall, and the relationship is approximately

$$F = 0.23U \tag{7.92}$$

and Eq. (7.91) becomes

$$H_R = 0.23AUaI \tag{7.93}$$

Values of a and I are given in engineering handbooks.

Mean Temperature Difference for Chilled-Water Cooling Coils

The heat transfer in chilled-water cooling coils depends on the **mean temperature difference** MTD between the air and the water in the tubes. This may be calculated from

$$\text{MTD} = \frac{(t_a - t_w') - (t_a' - t_w)}{\ln\left[(t_a - t_w')/(t_a' - t_w)\right]} \tag{7.94}$$

where MTD = mean temperature difference, °F
  t_a = temperature of entering air, °F
  t_a' = temperature of leaving air, °F
  t_w = temperature of entering water °F
  t_w' = temperature of leaving water, °F

The amount of surface is then determined by the formula

$$S = \frac{H}{K \times \text{MTD}} \tag{7.95}$$

where S = coil surface, ft^2
  H = *sensible* heat to be removed, Btu/h
  MTD = mean temperature difference, °F
  K = transmission coefficient, Btu/(h · ft^2 · °F)

The factor K depends upon the coil design and the water velocity.

The quantity of water circulated depends upon the total amount of heat removed from the air (sensible heat plus latent heat) and the temperature rise of the water which is to be allowed.

Cooling Tower Efficiency

The maximum cooling effect which could theoretically be obtained is limited by the wet-bulb temperature of the air. The efficiency of cooling is

$$E = \frac{t_1 - t_2}{t_1 - t} \quad (7.96)$$

where t = wet-bulb temperature of air
  t_1 = temperature of water entering tower
  t_2 = temperature of water leaving tower

The efficiency usually ranges between 70 and 75 percent. The water is usually cooled through a range of 10° and 15°, and its leaving temperature is obviously a function of the outside wet-bulb temperature.

When space is limited, a spray chamber, designed like a conventional air washer, is sometimes used for condensing-water cooling.

Some water is lost from a cooling tower as the result of evaporation and of drift. The total, for a system using a refrigerant such as F-12, is about 0.06 gal/(min · ton) of refrigeration. The amount of city water used and wasted, with no cooling tower, is from 1½ to 2 gal/(min · ton) depending upon the water temperature. The cooling tower is generally assumed to give a 90 to 95 percent saving of water.

USCS and SI Formulas for Air Conditioning and Heating

The formulas that follow are given first in USCS units and then in SI units. As such, the formulas allow easy comparison of the two systems of units. Abbreviations used in the formulas follow this presentation. The mechanical engineer can compute results in both systems of units and compare them to verify design results.

$$H_S = 1.08 \, \frac{\text{Btu} \cdot \text{min}}{\text{h} \cdot \text{ft}^3 \cdot °\text{F}} \times \text{CFM} \times \Delta T \quad (7.97)$$

$$H_{SM} = 72.42 \, \frac{\text{kJ} \cdot \text{min}}{\text{h} \cdot \text{m}^3 \cdot °\text{C}} \times \text{CMM} \times \Delta T_M \quad (7.98)$$

$$H_L = 0.68 \frac{\text{Btu} \cdot \text{min} \cdot \text{lb DA}}{\text{h} \cdot \text{ft}^3 \cdot \text{gr H}_2\text{O}} \times \text{CFM} \times \Delta W \tag{7.99}$$

$$H_{\text{LM}} = 177{,}734.8 \frac{\text{kJ} \cdot \text{min} \cdot \text{kg DA}}{\text{h} \cdot \text{m}^3 \cdot \text{kg H}_2\text{O}} \times \text{CMM} \times \Delta W_M \tag{7.100}$$

$$H_T = 4.5 \frac{\text{lb} \cdot \text{min}}{\text{h} \cdot \text{ft}^3} \times \text{CFM} \times \Delta h \tag{7.101}$$

$$H_{\text{TM}} = 72.09 \frac{\text{kg} \cdot \text{min}}{\text{h} \cdot \text{m}^3} \times \text{CMM} \times \Delta h_M \tag{7.102}$$

$$H_T = H_S + H_L \tag{7.103}$$

$$H_{\text{TM}} = H_{\text{SM}} + H_{\text{LM}} \tag{7.104}$$

$$H = 500 \frac{\text{Btu} \cdot \text{min}}{\text{h} \cdot \text{gal} \cdot {}^\circ\text{F}} \times \text{GPM} \times \Delta T \tag{7.105}$$

$$H_{\text{M}} = 250.8 \frac{\text{kJ} \cdot \text{min}}{\text{h} \cdot \text{L} \cdot {}^\circ\text{C}} \times \text{LPM} \times \Delta T_{\text{M}} \tag{7.106}$$

$$\frac{\text{AC}}{\text{HR}} = \frac{\text{CFM} \times 60 \text{ min/h}}{\text{VOLUME}} \tag{7.107}$$

$$\frac{\text{AC}}{\text{HR}_{\text{M}}} = \frac{\text{CMM} \times 60 \text{ min/h}}{\text{VOLUME}_{\text{M}}} \tag{7.108}$$

$$^\circ\text{C} = \frac{^\circ\text{F} - 32}{1.8} \tag{7.109}$$

$$^\circ\text{F} = 1.8{}^\circ\text{C} + 32 \tag{7.110}$$

where
H_S = sensible heat, Btu/h
H_{SM} = sensible heat, kJ/h
H_L = latent heat, Btu/h
H_{LM} = latent heat, kJ/h
H_T = total heat, Btu/h
H_{TM} = total heat, kJ/h
H = total heat, Btu/h
H_{M} = total heat, kJ/h
ΔT = temperature difference, °F
ΔT_{M} = temperature difference, °C
ΔW = humidity ratio difference, gr H$_2$O/lb DA
ΔW_{M} = humidity ratio difference, kg H$_2$O/kg DA
Δh = enthalpy difference, Btu/lb DA
Δh = enthalpy difference, kJ/lb DA

$$
\begin{aligned}
\text{CFM} &= \text{airflow rate, ft}^3/\text{min} \\
\text{CMM} &= \text{airflow rate, m}^3/\text{min} \\
\text{GPM} &= \text{water flow rate, gal/min} \\
\text{LPM} &= \text{water flow rate, L/min} \\
\text{AC/HR} &= \text{air change rate per hour, English} \\
\text{AC/HR}_M &= \text{air change rate per hour, SI} \\
\text{AC/HR} &= \text{AC/HR}_M \\
\text{VOLUME} &= \text{space volume, ft}^3 \\
\text{VOLUME}_M &= \text{space volume, m}^3 \\
\text{kJ/h} &= \text{Btu/h} \times 1.055 \\
\text{CMM} &= \text{CFM} \times 0.02832 \\
\text{LPM} &= \text{GPM} \times 3.785 \\
\text{kJ/lb} &= \text{Btu/lb} \times 2.326 \\
\text{m} &= \text{ft} \times 0.3048 \\
\text{m}^2 &= \text{ft}^2 \times 0.0929 \\
\text{m}^3 &= \text{ft}^3 \times 0.02832 \\
\text{kg} &= \text{lb} \times 0.4536 \\
1.0 \text{ GPM} &= 500 \text{ lb Steam/h} \\
1.0 \text{ lb Stm/h} &= 0.002 \text{ GPM} \\
1.0 \text{ lb H}_2\text{O/h} &= 1.0 \text{ lb Steam/h} \\
\text{kg/m}^3 &= \text{lb/ft}^3 \times 16.017 \text{ (density)} \\
\text{m}^3/\text{kg} &= \text{ft}^3/\text{lb} \times 0.0624 \text{ specific volume} \\
\text{kg H}_2\text{O/kg DA} &= \text{gr H}_2\text{O/lb DA}/7000 = \text{lb H}_2\text{O/lb DA}
\end{aligned}
$$

Alternative Steam Pipe Pressure Drop and Flow Rate Formulas

$$
\Delta P = \frac{0.01306 W^2 (1 + 3.6/\text{ID})}{3600 \times D \times \text{ID}^5} \tag{7.111}
$$

$$
W = 60 \sqrt{\frac{\Delta P \times D \times \text{ID}^5}{0.01306 \times (1 + 3.6/\text{ID})}} \tag{7.112}
$$

$$
W = 0.41667 V A_{\text{INCHES}} D = 60 V A_{\text{FEET}} D \tag{7.113}
$$

$$
V = \frac{2.4W}{A_{\text{INCHES}} D} = \frac{W}{60 A_{\text{FEET}} D} \tag{7.114}
$$

where ΔP = pressure drop per 100 ft of pipe (psig/100 ft)
$\quad\quad W$ = steam flow rate, lb/h
$\quad\quad \text{ID}$ = actual inside diameter of pipe, in
$\quad\quad D$ = average density of steam at system pressure, lb/ft^3
$\quad\quad V$ = velocity of steam in pipe, ft/min
$\quad A_{\text{INCHES}}$ = actual cross-sectional area of pipe, in^2
$\quad A_{\text{FEET}}$ = actual cross-sectional area of pipe, ft^2

Condensate Piping Formulas

$$\text{FS} = \frac{H_{\text{SSS}} - H_{\text{SCR}}}{H_{\text{LCR}}} \times 100 \qquad (7.115)$$

$$W_{\text{CR}} = \frac{\text{FS}}{100} \times W \qquad (7.116)$$

where FS = flash steam, %
H_{SSS} = sensible heat at steam supply pressure, Btu/lb
H_{SCR} = sensible heat at condensate return pressure, Btu/lb
H_{LCR} = latent heat at condensate return pressure, Btu/lb
W = steam flow rate, lb/h
W_{CR} = condensate flow based on percentage of flash steam created during condensing process, lb/h. Use this flow rate in steam equations above to determine condensate return pipe size

HVAC Efficiency Formulas

$$\text{COP} = \frac{\text{BTU OUTPUT}}{\text{BTU INPUT}} = \frac{\text{EER}}{3.413} \qquad (7.117)$$

$$\text{EER} = \frac{\text{BTU OUTPUT}}{\text{WATTS INPUT}} \qquad (7.118)$$

Turndown ratio = maximum firing rate: minimum firing rate

(that is, 5:1, 10:1, 25:1)

$$\text{OVERALL THERMAL EFF} = \frac{\text{GROSS BTU OUTPUT}}{\text{GROSS BTU INPUT}}$$
$$\times 100\% \qquad (7.119)$$

$$\text{COMBUSTION EFF} = \frac{\text{BTU INPUT} - \text{BTU STACK LOSS}}{\text{BTU INPUT}}$$
$$\times 100\% \qquad (7.120)$$

Overall thermal efficiency range   75%–90%
Combuston efficiency range   85%–95%

Formulas for HVAC Equipment Room Ventilation

For completely enclosed equipment rooms:

$$\text{CFM} = 100 \times G^{0.5} \qquad (7.121)$$

where CFM = exhaust airflow rate required, ft^3/min
G = mass of refrigerant of largest system, lb

For partially enclosed equipment rooms:

$$\text{FA} = G^{0.5} \qquad (7.122)$$

where FA = ventilation-free opening area, ft^2
G = mass of refrigerant of largest system, lb

PSYCHROMETRIC FORMULAS

The following formulas are from Carrier Corporation publications,† and they cover air mixing, cooling loads, sensible heat factor, bypass factor, temperature at the apparatus, supply air temperature, air quantity, and derivation of air constants. Abbreviations and symbols for the formulas are given below.

Abbreviations

adp	apparatus dew point
BF	bypass factor
(BF) (OALH)	bypassed outdoor air latent heat
(BF) (OASH)	bypassed outdoor air sensible heat
(BF) (OATH)	bypassed outdoor air total heat
Btu/h	British thermal units per hour
cfm, ft^3/min	cubic feet per minute
db	dry-bulb
dp	dew point
ERLH	effective room latent heat
ERSH	effective room sensible heat
ERTH	effective room total heat
ESHF	effective sensible heat factor

---

† *Handbook of Air-Conditioning System Design*, McGraw-Hill, New York, various dates.

°F degrees Fahrenheit

fpm, ft/min feet per minute

gpm, gal/min gallons per minute

gr/lb grains per pound

GSHF grand sensible heat factor

GTH grand total heat

GTHS grand total heat supplement

OALH outdoor air latent heat

OASH outdoor air sensible heat

OATH outdoor air total heat

rh relative humidity

RLH room latent heat

RLHS room latent heat supplement

RSH room sensible heat

RSHF room sensible heat factor

RSHS room sensible heat supplement

RTH room total heat

Sat Eff saturation efficiency of sprays

SHF sensible heat factor

TLH total latent heat

TSH total sensible heat

wb wet-bulb

Symbols

cfm_{ba} bypassed air quantity around apparatus

cfm_{da} dehumidified air quantity

cfm_{oa} outdoor air quantity

cfm_{ra} return air quantity

cfm_{sa} supply air quantity

h specific enthalpy

h_{adp} apparatus dew point enthalpy

h_{cs} effective surface temperature enthalpy

h_{ea} entering air enthalpy

h_{la} leaving air enthalpy

h_m mixture of outdoor and return air enthalpy

h_{oa} outdoor air enthalpy

h_{rm} room air enthalpy

h_{sa} supply air enthalpy

t temperature

t_{adp} apparatus dew point temperature

t_{edb} entering dry-bulb temperature

t_{es} effective surface temperature

t_{ew} entering water temperature

t_{ewb} entering wet-bulb temperature

t_{ldb} leaving dry-bulb temperature

t_{lw} leaving water temperature

t_{lwb} leaving wet-bulb temperature

t_m mixture of outdoor and return air dry-bulb temperature

t_{oa} outdoor air dry-bulb temperature

t_{rm} room dry-bulb temperature

t_{sa} supply air dry-bulb temperature

W moisture content or specific humidity

W_{adp} apparatus dew point moisture content

W_{ea} entering air moisture content

W_{es} effective surface temperature moisture content

W_{la} leaving air moisture content

W_m mixture of outdoor and return air moisture content

W_{oa} outdoor air moisture content

W_{rm} room moisture content

W_{sa} supply air moisture content

Air Mixing Formulas (Outdoor and Return Air)

$$t_m = \frac{\text{cfm}_{oa} \times t_{oa} + \text{cfm}_{ra} \times t_{rm}}{\text{cfm}_{sa}} \qquad (7.123)$$

$$h_m = \frac{(\text{cfm}_{oa} \times h_{oa}) + (\text{cfm}_{ra} \times h_{rm})}{\text{cfm}_{sa}} \qquad (7.124)$$

$$W_m = \frac{(\text{cfm}_{oa} \times W_{oa}) + (\text{cfm}_{ra} \times W_{rm})}{\text{cfm}_{sa}} \qquad (7.125)$$

Cooling Load Formulas

$$\text{ERSH} = \text{RSH} + (\text{BF})(\text{OASH}) + \text{RSHS}\dagger \tag{7.126}$$

$$\text{ERLH} = \text{RLH} + (\text{BF})(\text{OALH}) + \text{RLHS}\dagger \tag{7.127}$$

$$\text{ERTH} = \text{ERLH} + \text{ERSH} \tag{7.128}$$

$$\text{TSH} = \text{RSH} + \text{OASH} + \text{RSHS}\dagger \tag{7.129}$$

$$\text{TLH} = \text{RLH} + \text{OALH} + \text{RLHS}\dagger \tag{7.130}$$

$$\text{GTH} = \text{TSH} + \text{TLH} + \text{GTHS}\dagger \tag{7.131}$$

$$\text{RSH} = 1.08\ddagger \times \text{cfm}_{sa} \times (t_{rm} - t_{sa}) \tag{7.132}$$

$$\text{RLH} = 0.68\ddagger \times \text{cfm}_{sa} \times (W_{rm} - W_{sa}) \tag{7.133}$$

$$\text{RTH} = 4.45\ddagger \times \text{cfm}_{sa} \times (h_{rm} - h_{sa}) \tag{7.134}$$

$$\text{RTH} = \text{RSH} + \text{RLH} \tag{7.135}$$

$$\text{OASH} = 1.08 \times \text{cfm}_{oa}(t_{oa} - t_{rm}) \tag{7.136}$$

$$\text{OALH} = 0.68 \times \text{cfm}_{oa}(W_{oa} - W_{rm}) \tag{7.137}$$

$$\text{OATH} = 4.45 \times \text{cfm}_{oa}(h_{oa} - h_{rm}) \tag{7.138}$$

$$\text{OATH} = \text{OASH} + \text{OALH} \tag{7.139}$$

$$(\text{BF})(\text{OATH}) = (\text{BF})(\text{OASH}) + (\text{BF})(\text{OALH}) \tag{7.140}$$

$$\text{ERSH} = 1.08 \times \text{cfm}_{da}\S \times (t_{rm} - t_{adp})(1 - \text{BF}) \tag{7.141}$$

$$\text{ERLH} = 0.68 \times \text{cfm}_{da}\S \times (W_{rm} - W_{adp})(1 - \text{BF}) \tag{7.142}$$

$$\text{ERTH} = 4.45 \times \text{cfm}_{da}\S \times (h_{rm} - h_{adp})(1 - \text{BF}) \tag{7.143}$$

$$\text{TSH} = 1.08 \times \text{cfm}_{da}\S \times (t_{edb} - t_{ldb})\dagger \tag{7.144}$$

$$\text{TLH} = 0.68 \times \text{cfm}_{da}\S \times (W_{ea} - W_{la})\dagger \tag{7.145}$$

$$\text{GTH} = 4.45 \times \text{cfm}_{da}\S \times (h_{ea} - h_{la})\dagger \tag{7.146}$$

---

† RSHS, RLHS and GTHS are supplementary loads due to duct heat gain, duct leakage loss, fan and pump horsepower gains, etc.

‡ See below for the derivation of these air constants.

§ When no air is to be physically bypassed around the conditioning apparatus, $\text{cfm}_{da} = \text{cfm}_{sa}$.

Sensible Heat Factor Formulas

$$\text{RSHF} = \frac{\text{RSH}}{\text{RSH} + \text{RLH}} = \frac{\text{RSH}}{\text{RTH}} \tag{7.147}$$

$$\text{ESHF} = \frac{\text{ERSH}}{\text{ERSH} + \text{ERLH}} = \frac{\text{ERSH}}{\text{ERTH}} \tag{7.148}$$

$$\text{GSHF} = \frac{\text{TSH}}{\text{TSH} + \text{TLH}} = \frac{\text{TSH}}{\text{GTH}} \tag{7.149}$$

Bypass Factor Formulas

$$\text{BF} = \frac{t_{\text{ldb}} - t_{\text{adp}}}{t_{\text{edb}} - t_{\text{adp}}} \qquad 1 - \text{BF} = \frac{t_{\text{edb}} - t_{\text{ldb}}}{t_{\text{edb}} - t_{\text{adp}}} \tag{7.150}$$

$$\text{BF} = \frac{W_{\text{la}} - W_{\text{adp}}}{W_{\text{ea}} - W_{\text{adp}}} \qquad 1 - \text{BF} = \frac{W_{\text{ea}} - W_{\text{la}}}{W_{\text{ea}} - W_{\text{adp}}} \tag{7.151}$$

$$\text{BF} = \frac{h_{\text{la}} - h_{\text{adp}}}{h_{\text{ea}} - h_{\text{adp}}} \qquad 1 - \text{BF} = \frac{h_{\text{ea}} - h_{\text{la}}}{h_{\text{ea}} - h_{\text{adp}}} \tag{7.152}$$

Temperature Formulas at Apparatus

$$t_{\text{edb}}\dagger = \frac{(\text{cfm}_{\text{oa}} \times t_{\text{oa}}) + (\text{cfm}_{\text{ra}} \times t_{\text{rm}})}{\text{cfm}_{\text{sa}}\S} \tag{7.153}$$

$$t_{\text{ldb}} = t_{\text{adp}} + \text{BF}(t_{\text{edb}} - t_{\text{adp}}) \tag{7.154}$$

Both t_{ewb} and t_{lwb} correspond to the calculated values of h_{ea} and h_{la} on the psychrometric chart.

$$h_{\text{ea}}\dagger = \frac{(\text{cfm}_{\text{oa}} \times h_{\text{oa}}) + (\text{cfm}_{\text{ra}} \times h_{\text{rm}})}{\text{cfm}_{\text{sa}}\S} \tag{7.155}$$

$$h_{\text{la}} = h_{\text{adp}} + \text{BF}(h_{\text{ea}} - h_{\text{adp}}) \tag{7.156}$$

Temperature Formulas for Supply Air

$$t_{\text{sa}} = t_{\text{rm}} - \frac{\text{RSH}}{1.08\text{cfm}_{\text{sa}}\S} \tag{7.157}$$

---

† When t_m, W_m, and h_m are equal to the entering conditions at the cooling apparatus, they may be substituted for t_{edb}, W_{ea}, and h_{ea}, respectively.
§ See footnote on page 173.

Air Quantity Formulas

$$\text{cfm}_{da} = \frac{\text{ERSH}}{1.08(1 - \text{BF})(t_{rm} - t_{adp})} \tag{7.158}$$

$$\text{cfm}_{da} = \frac{\text{ERLH}}{0.68(1 - \text{BF})(W_{rm} - W_{adp})} \tag{7.159}$$

$$\text{cfm}_{da} = \frac{\text{ERTH}}{4.45(1 - \text{BF})(h_{rm} - h_{adp})} \tag{7.160}$$

$$\text{cfm}_{da}\S = \frac{\text{TSH}}{1.08(t_{edb} - t_{ldb})} \tag{7.161}$$

$$\text{cfm}_{da}\S = \frac{\text{TLH}}{0.68(W_{ea} - W_{la})} \tag{7.162}$$

$$\text{cfm}_{da}\S = \frac{\text{GTH}}{4.45(h_{ea} - h_{la})} \tag{7.163}$$

$$\text{cfm}_{sa} = \frac{\text{RSH}}{1.08(t_{rm} - t_{sa})} \tag{7.164}$$

$$\text{cfm}_{sa} = \frac{\text{RLH}}{0.68(W_{rm} - W_{sa})} \tag{7.165}$$

$$\text{cfm}_{sa} = \frac{\text{RTH}}{4.45(h_{rm} - h_{sa})} \tag{7.166}$$

$$\text{cfm}_{ba} = \text{cfm}_{sa} - \text{cfm}_{da} \tag{7.167}$$

Note: cfm_{da} will be less than cfm_{sa} only when air is physically bypassed around the conditioning apparatus.

$$\text{cfm}_{sa} = \text{cfm}_{oa} + \text{cfm}_{ra} \tag{7.168}$$

Derivation of Air Constants

$$1.08 = 0.244 \times \frac{60}{13.5}$$

where 0.244 = specific heat of moist air at 70°F db and 50% rh, Btu/
(°F · lb DA)
    60 = min/h
  13.5 = specific volume of moist air at 70°F db and 50% rh

---

§ See footnote on page 173.

$$0.68 = \frac{60}{13.5} \times \frac{1076}{7000}$$

where 60 = min/h
13.5 = specific volume of moist air at 70°F db and 50% rh
1076 = average heat removal required to condense 1 lb water vapor from the room air
7000 = gr/lb

$$4.45 = \frac{60}{13.5}$$

where 60 = min/h
13.5 = specific volume of moist air at 70°F db and 50% rh

FORMULAS FOR STEAM TRAP CHOICE

The selection of the trap for the steam mains or risers is dependent on the pipe warm-up load and the radiation load from the pipe. Warm-up load is the condensate which is formed by heating the pipe surface when the steam is first turned on. For practical purposes, the final temperature of the pipe is the steam temperature. Warm-up load is determined from

$$C_1 = \frac{W(t_f - t_i)(0.114)}{h_l T} \qquad (7.169)$$

where C_1 = warm-up condensate, lb/h
W = total weight of pipe, lb (from tables in engineering handbooks)
t_f = final pipe temperature, °F (steam temp.)
t_i = initial pipe temperature, °F (usually room temp.)
0.114 = specific heat constant for wrought iron or steel pipe (0.092 for copper tubing)
h_l = latent heat of steam, Btu/lb (from steam tables)
T = time for warm-up, h

The radiation load is the condensate formed by unavoidable radiation loss from a bare pipe. This load is determined from the following equation and is based on still air surrounding the steam main or riser:

$$C_2 = \frac{LK(t_f - t_i)}{h_l} \qquad (7.170)$$

where C_2 = radiation condensate, lb/h
L = linear length of pipe, ft
K = heat transmission coefficient, Btu/(h · lin ft · °F)

The radiation load builds up as the warm-up load drops off under normal operating conditions. The peak occurs at the midpoint of the warm-up cycle. Therefore, one-half of the radiation load is added to the warm-up load to determine the amount of condensate that the trap handles.

Safety Factor

Good design practice dictates the use of safety factors in steam trap selection. Safety factors from 2 to 1 to as high as 8 to 1 may be required, and for the following reasons:

1. The steam pressure at the trap inlet or the back pressure at the trap discharge may vary. This changes the steam trap capacity.
2. If the trap is sized for normal operating load, condensate may back up into the steam lines or apparatus during start-up or warm-up operation.
3. If the steam trap is selected to discharge a full and continuous stream of water, the air could not be vented from the system.

The following guide is used to determine the safety factor:

Design	Safety Factor
Draining steam main	3 to 1
Draining steam riser	2 to 1
Between boiler and end of main	2 to 1
Before reducing valve	3 to 1
Before shutoff valve (closed part of time)	3 to 1
Draining coils	3 to 1
Draining apparatus	3 to 1

When the steam trap is to be used in a high-pressure system, determine whether the system is to operate under low-pressure conditions at certain intervals such as nighttime or weekends. If this condition is likely to occur, then an additional safety factor should be considered to account for the lower pressure drop available during nighttime operation.

SECTION 8
THERMODYNAMICS FORMULAS

NOMENCLATURE

a = area of piston (with deductions for piston and tail rods, when present), in^2 (mm^2)

C = constant = 12 for four-cycle engine, 20 for two-cycle engine

C_p = heat capacity at constant pressure, Btu/(lb·°F) [kJ/(kg·°C)]

C_v = heat capacity at constant volume, Btu/(lb·°F) [kJ/(kg·°C)]

c = clearance, decimal fraction of displacement

D = piston displacement, ft^3/cycle (m^3/cycle)

D' = piston displacement, ft^3/min (m^3/m)

d = bore, in (mm)

F = net force at arm bearing point, lb (N)

g_c = conversion factor = 32.2 lbm·ft/(lbf·s^2) (9.81 m/s^2)

h = enthalpy (= $pv/J + u$, Btu/lb (kJ/kg)

h_s = static head, ft (m) fluid

h_t = total head, ft (m) fluid

h_v = velocity head, ft (m) fluid

h_w'' = head, inH$_2$O (mmH$_2$O)

h_1 = enthalpy at compressor supply, Btu/lb (kJ/kg)

h_2 = enthalpy at compressor delivery, Btu/lb (kJ/kg)

J = mechanical equivalent of heat, 778 ft·lb/Btu

k = ratio of specific heats = C_p/C_v

L = length of stroke, ft (m)

$\quad\quad$ = length of brake arm (shaft centerline to arm bearing point), ft (m)

l = stroke length, in

m = molecular weight, lb (kg)

$\dot{m}$ = number of lb mol (kg mol)

mep = mean effective pressure, psi (kPa)

N = shaft rotation speed, rpm

$\quad$ = number of stages

n = number cycles completed per minute

$\quad$ = polytropic exponent, pv^n = constant for polytropic process

p = pressure, lbf/ft² abs (kPa)

p_1 = supply pressure, lbf/in² abs (kPa)

p_2 = delivery pressure, lbf/in² abs (kPa)

$\Delta p = p_c - p_b$, pressure rise across pump, lbf/in² (kPa)

Q = volumetric flow rate, ft³/min (m³/min)

$\quad$ = heat recoverable, Btu/h (kJ/h)

Q_1 = heat added, Btu/lb (kJ/kg)

Q_2 = heat rejected, Btu/lb (kJ/kg)

ΔQ = heat added or abstracted, Btu/lb (kJ/kg)

R = gas constant, characteristic for gas or mixture

R' = universal gas constant = 1544 ft · lb/(lb · °R)

R_p = ratio of pressures = p_2/p_1

R_{p1}, R_{p2}, R_{p3} = compression ratio in stages 1, 2, 3, respectively

R_t = ratio of temperatures = T_2/T_1

R_v = ratio of volumes = v_1/v_2

S = entropy, Btu/(lb · °F) [kJ/(kg · °C)]

shp = shaft horsepower of engine, hp (kW)

T = absolute temperature

T_1 = temperature of heat addition, °R

T_2 = temperature of heat rejection, °R

ΔT = temperature drop of exhaust gases through heat exchanger, °F (°C)

u = internal or intrinsic energy, Btu/lb (kJ/kg)

du = change in internal energy, Btu/lb (kJ/kg)

$\overline{V}$ = velocity, ft/s (m/s)

v = volume, ft³ (m³)

$\overline{v}$ = specific volume, ft³/lb (m³/kg)

w = weight, lb (kg)

$\overline{w}_f$ = density of fluid, lb/ft³ (kg/m³)

$\overline{w}_w$ = density of water, 62.4 lb/ft³ (998.4 kg/m³)

dw = work, ft·lbf (N·m)

ΔW = work done on or by fluid, ft·lbf/lb (N·m/kg)

Z = elevational energy, ft·lbf/lbm (N·m/kg mass)

$_1$ = entering or initial conditions

$_2$ = leaving or final conditions

GENERAL ENERGY EQUATION

The general energy equation for fluid moving through a system is

$$Z_1 + \frac{\overline{V}_1^2}{2g_c} + p_1\overline{v}_1 + u_1 J = Z_2 + \frac{\overline{V}_2^2}{2g_c} + p_2\overline{v}_2 + u_2 J \pm \Delta W \pm \Delta QJ \tag{8.1}$$

or

$$Z_1 + \frac{\overline{V}_1^2}{2g_c} + h_1 J = Z_2 + \frac{\overline{V}_2^2}{2g_c} + h_2 J \pm \Delta W \pm \Delta QJ \tag{8.2}$$

where all items must be in a consistent system of units (Fig. 8.1).

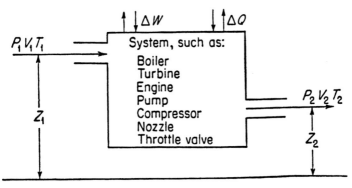

FIGURE 8.1

GAS LAWS

For ideal, or perfect, gases, Boyle's law, at constant temperature, is

$$p_1 v_1 = p_2 v_2 = \text{const} \tag{8.3}$$

Charles' law, at constant pressure, is

$$\frac{v_1}{T_1} = \frac{v_2}{T_2} \tag{8.4}$$

Gay-Lussac's law, at constant volume, is

$$\frac{p_1}{T_1} = \frac{p_2}{T_2} \tag{8.5}$$

Combined Boyle's, Charles', and Gay-Lussac's Laws   If both pressure and temperature change simultaneously, then the volume is

$$v_2 = v_1 \frac{p_1}{p_2} \frac{T_2}{T_1} \tag{8.6}$$

Characteristic Gas Equation   The combination of Boyle's and Charles' laws gives the characteristic equation of a perfect gas

$$pv = wRT \tag{8.7}$$

Values of R for various gases are given in engineering handbooks.

By Avogadro's hypothesis, equal volumes of gases at the same pressure and temperature contain the same number of molecules, or the molecular weight in pounds occupies the same volume (358.7 ft^3 at 29.92 inHg and $32°F$). This fact can be introduced into Eq. (8.7) to give the molal form of the characteristic gas equation

$$pv = \overline{m} R' T \tag{8.8}$$

Although Eq. (8.8) does not rigorously apply to actual gases, it can be used with adequate accuracy in many cases. A correction factor to account for nonideal behavior may be introduced into the right-hand side of Eq. (8.8) as a multiplying device.

Nonflow Processes or Phases for Perfect Gases

If a quantity of perfect gas is trapped in a body or system, there are five clearly identifiable ways in which energy can be added to or abstracted from the system: isothermal, isentropic, constant-volume, constant-pressure, and polytropic. These processes or phases are represented in Fig. 8.2. The equa-

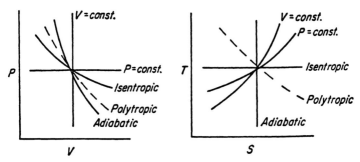

FIGURE 8.2

tions for evaluating the changes in the properties and the performances are given in Table 8.1 for an ideal gas. The evaluation of the energy addition is accomplished by the simple energy equation

$$dQ = du + dw \qquad (8.9)$$

In the calculation of the pressure, volume, and temperature changes with various values for the exponents n and k, it is good to recognize that ratios and ratios raised to exponential powers can be used as divisors or multipliers to facilitate the finding of numerical answers. Figure 8.3 is helpful in this respect.

COMPRESSOR PERFORMANCE

Adiabatic (Isentropic) Compressor Standards

The ideal compressor cycle is shown in Fig. 8.4 where there are three phases: (1) admission from a to 1, (2) compression from 1 to 2, and (3) delivery from 2 to b. For a perfect gas with reversible adiabatic (or isentropic) compression ($pv^k = $ const) the work is given

$$\Delta W_{\text{adiabatic cycle}} = 144 p_1 v_1 \left(\frac{k}{k-1} \right) (R_p^{(k-1)/k} - 1) \qquad (8.10)$$

If the compression is isentropic and for a real gas whose thermodynamic properties are known (as for refrigerants), then for a perfect compressor the work of the cycle is

$$\Delta W_{\text{adiabatic cycle}} = 778(h_2 - h_1) \qquad (8.11)$$

Equations (8.10) and (8.11) give identical answers for a perfect gas.

TABLE 8.1 Relationships for Ideal Gases

Process	p,v,T relation	Nonflow work $\int_1^2 P\,dv$	Q	Δu	Δh	ΔS
Isothermal ($T = $ const)	$p_1 v_1 = p_2 v_2$	$p_1 v_1 \ln R_v$	$\dfrac{p_1 v_1}{J} \ln R_v$	0	0	$\dfrac{wR}{J} \ln R_v$
Constant pressure $p = $ const	$\dfrac{T_2}{T_1} = \dfrac{v_2}{v_1}$	$p(v_2 - v_1)$	$wC_p(T_2 - T_1)$	$wC_v(T_2 - T_1)$	$wC_p(T_2 - T_1)$	$wC_p \ln R_t$
Constant volume ($v = $ const)	$\dfrac{T_2}{T_1} = \dfrac{p_2}{p_1}$	0	$wC_v(T_2 - T_1)$	$wC_v(T_2 - T_1)$	$wC_p(T_2 - T_1)$	$wC_v \ln R_t$
Isentropic ($S = $ const)	$(p_1 v_1)^k = (p_2 v_2)^k$ $\dfrac{T_1}{T_2} = R_v^{k-1}$ $= \left(\dfrac{1}{R_p}\right)^{(k-1)/k}$	$\dfrac{p_2 v_2 - p_1 v_1}{1 - k}$	0	$wC_v(T_2 - T_1)$	$wC_p(T_2 - T_1)$	0
Polytropic ($pv^n = $ const)	$(p_1 v_1)^n = (p_2 v_2)^n$ $\dfrac{T_1}{T_2} = R_v^{n-1}$ $= \left(\dfrac{1}{R_p}\right)^{(n-1)/n}$	$\dfrac{p_2 v_2 - p_1 v_1}{1 - n}$	$wC_v \dfrac{k - n}{1 - n}(T_2 - T_1)$	$wC_v(T_2 - T_1)$	$wC_p(T_2 - T_1)$	$wC_v \dfrac{k - n}{1 - n} \ln R_t$

v_2 T_2 p_2

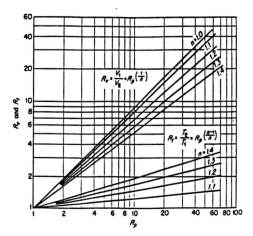

FIGURE 8.3

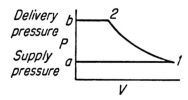

FIGURE 8.4

Isothermal Compressor Standards

If compressors are so cooled that temperature is constant during compression, then the isothermal standard prevails, as shown in Fig. 8.5 ($pv =$ const) and there is a saving in work over the adiabatic value of Eq. (8.10). The work is given as

$$\Delta W_{\text{isothermal}} = 144 p_1 v_1 \ln R_p \qquad (8.12)$$

Multistage Compressor Standards

The work of a compressor is reduced by the use of multistage compression with intercooling between stages. If cooling is complete and the gas enters the succeeding stage at the same temperature at which it enters the machine, the intercooling is said to be *perfect*. Minimum work is then obtained with unique values of pressure between the stages, called *best receiver pressure*. It is determined by

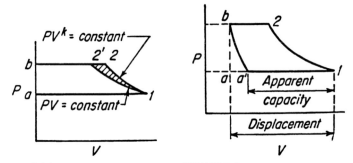

FIGURE 8.5   **FIGURE 8.6**

$$R_{p1} = R_{p2} = R_{p3} = \cdots = R_p^{1/N} \qquad (8.13)$$

With isentropic compression in each stage, best receiver pressure, and perfect intercooling, the work of the ideal cycle is

$$\Delta W_{\text{multistage}} = 144 N p_1 v_1 \left(\frac{k}{k-1}\right)(R_p^{(k-1)/Nk} - 1) \qquad (8.14)$$

The isothermal standard of Eq. (8.12) applies equally well to single- and multistage compression.

Capacity

Capacity is expressed on a volume basis and for air is given on the "free air" basis. This measures capacity at the ambient pressure, temperature, and humidity.

For a positive-displacement machine without clearance, the volume is represented as $v_1 - v_a$ in Fig. 8.6. This is obtained from the dimensions of the cylinders.

$$\text{Piston displacement, ft}^3/\text{cycle} = D = \left(\frac{\pi d^2}{4}\right)\left(\frac{l}{1728}\right) \qquad (8.15)$$

$$D' = \text{piston displacement, ft}^3/\text{min} = \frac{\pi d^2}{4} nl \left(\frac{1}{1728}\right)$$

$$= \frac{d^2 ln}{2200} \qquad (8.16)$$

As the machine contains clearance which runs from 2 to 20 percent of the displacement, there is a clearance reexpansion loss, so that point a shifts to position a' and the length $v_1 = v_a'$ is less than the displacement $v_1 - v_a$ in Fig. 8.6. This apparent capacity is calculable by

$$\text{Apparent capacity} = D(1 + c - cR_p^{1/k}) \tag{8.17}$$

The actual capacity, as metered, for a real compressor is less than this apparent value because of suction heating, suction pressure drop, and leakage losses.

Volumetric efficiency is the ratio of capacity to displacement, and if the former is a real metered value, then

$$\text{Actual volumetric eff, } \% = \frac{\text{actual metered capacity}}{\text{piston displacement}} \times 100 \tag{8.18}$$

If the apparent capacity is used from Eq. (8.17), then

$$\text{Apparent volumetric eff, } \% = (1 + c - cR_p^{1/k}) \times 100 \tag{8.19}$$

The relation between these two is called the *slippage efficiency* and is defined as

$$\text{Slippage eff, } \% = \frac{\text{actual volumetric eff}}{\text{apparent volumetric eff}} \times 100 \tag{8.20}$$

Ideal Horsepower (kW) of Compressors

The equations for ideal work—(8.10), (8.11), (8.12), and (8.14)—apply equally well to compressors involving clearance, because work is independent of clearance. For a volume flow rate of 100 ft³/min (m³/min), the equations can be rewritten as

$$\frac{\text{Isentropic or adiabatic hp}}{100 \text{ ft}^3/\text{min}} = \frac{k}{k-1} \frac{p_1}{2.292} (R_p^{(k-1)/k} - 1) \tag{8.21}$$

$$= \frac{(h_2 - h_1) \text{ lb/min}}{0.4242\bar{v}_1} \tag{8.22}$$

where $\bar{v}_1$ = specific volume, ft³/lb, at supply pressure

$$\frac{\text{Isothermal hp}}{100 \text{ ft}^3/\text{min}} = \frac{p_1}{2.292} \ln R_p \tag{8.23}$$

In a multistage compressor with perfect intercooling and best receiver pressure,

$$\frac{\text{Isentropic or adiabatic hp}}{100 \text{ ft}^3/\text{min}} = \frac{Np_1}{2.292} \frac{k}{k-1} (R_p^{(k-1)/Nk} - 1) \tag{8.24}$$

Compression Efficiency

The actual power required by a compressor can be compared to the ideal power (for the same capacity) to give

$$\text{Compression eff, \%} = \frac{\text{ideal hp}}{\text{actual hp}} \times 100 \qquad (8.25)$$

The ideal value may be obtained from Eqs. (8.21) to (8.24) which give adiabatic and isothermal compression efficiencies.

The actual horsepower (kW) may be obtained from the compressor-cylinder indicator card or the shaft of the compressor, or it may be actual power input to the motor terminals of an electrically driven unit. Care must be taken to specify the base.

FAN PERFORMANCE

A fan is a compressor in which the change in density of the gaseous fluid, on passage through the machine, is negligibly small.

Definitions

Standard air: Air at 68°F, 29.92 inHG pressure, and 5 percent relative humidity. It has a density of 0.07488 lb/ft³ and a specific volume of 13.3 ft³/lb and is the basis for measuring fan performance.

Capacity: The volume Q delivered by a fan, expressed in cubic feet (meters) per minute.

Head: The difference between the pressures on the suction and discharge sides of a fan, variously expressed as feet (meters) of fluid, inches (millimeters) of water, pounds per square inch h_w'' (kilopascals), etc. Conversion is as follows

$$h_t = \frac{h_w''}{12} \frac{\overline{w}_w}{w_f} \qquad \text{ft (m) fluid} \qquad (8.26)$$

$$h_t = 69.5 h_w'' \qquad \text{ft (m) std air} \qquad (8.27)$$

$$\text{Pressure} = \frac{h_w''}{27.7} \qquad \text{psi (kPa)} \qquad (8.28)$$

Static, Velocity, and Total Heads

As shown in Fig. 8.7, three pressures can be read in a fan duct. *Static head* h_s is a directly obtained pressure reading, and *velocity head* h_v is obtained

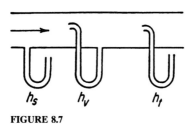

FIGURE 8.7

from the flow rate in the duct and must be an average value obtained by a traverse. Then *total head* is

$$h_t = h_s + h_v \qquad (8.29)$$

Conversion is done by

$$\text{Velocity} = \sqrt{2g_c h_v} \qquad \text{ft/s (m/s)} \qquad (8.30)$$

Substituting the conversion of Eq. (8.26) gives

$$\text{Velocity} = 1096.2 \sqrt{\frac{h_w''}{w_f}} \qquad \text{ft/min (m/min)} \qquad (8.31)$$

and if standard air is used,

$$\text{Velocity} = 4005 \sqrt{h_w''} \qquad \text{ft/min (m/min)} \qquad (8.32)$$

Fan performance is variously based on static head and total head, the former being generally more realistic because it is the only form of head usable in overcoming a system resistance.

Horsepower of Fans

The ideal or air horsepower is given by

$$\text{Air hp (kW)} = \frac{Q h_w''}{6355} \qquad (8.33)$$

Static or total head may be used, giving two alternative values of ideal horsepower, the latter being larger.

Shaft Horsepower (shp)

Shaft horsepower input to drive the fan is measured by a suitable dynamometer. *Fan efficiency* is defined as

$$\text{Fan eff, \%} = \frac{\text{air hp}}{\text{shp}} \times 100 \qquad (8.34)$$

This value may be on the static or total basis.

Fan Characteristics

Fans, like other fluid acceleration machines, operate with characteristic curves. A set of characteristics is plotted in Fig. 8.8 for a representative fan. These curves are exactly definitive, and the fan must operate at some point on the characteristic.

Fan Laws

See Sec. 7 for fan laws.

PERFORMANCE CHARACTERISTICS OF PISTON MACHINES

Mean Effective Pressure

In piston and cylinder machines, it is convenient to measure performance through the use of mean effective pressure. As illustrated by Fig. 8.9, the *mean effective pressure* is defined as the difference in pressure between the two sides of the piston, which difference tends to move the piston in an engine or resist its motion in a pump. It can be defined as

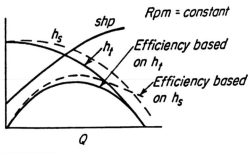

FIGURE 8.8

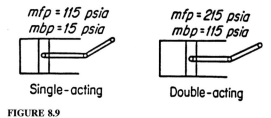

FIGURE 8.9

Mean effective pressure (mep) = mean forward pressure (mfp)

$$- \text{mean back pressure (mbp)} \quad (8.35)$$

Thus in the two illustrations of Fig. 8.9, for a single-acting and a double-acting mechanism, the mep is the same (100 psi) because in *A*,

$$\text{mep} = 115 - 15 = 100 \text{ psi}$$

and in *B*

$$\text{mep} = 215 - 115 = 100 \text{ psi (kPa)}$$

These pressures are mean values which can be considered as prevailing throughout the stroke. They may be calculated for ideal cyclic conditions by utilizing the methods of thermodynamics and fluid dynamics. Thus in Fig. 8.10 the area of the *P-V* diagram is the work of the cycle, expressible in foot-pounds. If that area is divided by the length of the diagram, i.e., by the stroke or displacement, the result is a vertical height for a rectangle of equivalent area, and this height is the mep, as

$$\text{mep} = \frac{\text{work of cycle, ft} \cdot \text{lb}}{\text{displacement, ft}^3 \times 144 \text{ in}^2/\text{ft}^2} = \frac{\text{lb}}{\text{in}^2} \quad (8.36)$$

The actual mean effective pressure is determined from the planimetered

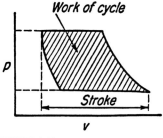

FIGURE 8.10

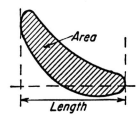

FIGURE 8.11

area of the indicator card, dividing by length and applying the spring scale. Thus with Fig. 8.11,

$$\text{mep} = \frac{\text{area, in}^2}{\text{length in}} \times \text{spring scale, lb/(in}^2 \cdot \text{in)} \tag{8.37}$$

Indicated Horsepower

Mean effective pressure is used to calculate the indicated horsepower. Thus

$$\text{Indicated hp} = \text{mep} \frac{Lan}{33,000} \tag{8.38}$$

The items a and L are obtained directly from the bore and stroke. The number of cycles completed per minute depends upon the mechanism construction, i.e., single- or double-acting, number of cylinders, and number of strokes or revolutions needed to complete a cycle.

Brake or Shaft Horsepower

As measured on a prony brake or dynamometer, brake or shaft horsepower is determined by

$$\text{hp} = \frac{2\pi LFN}{33,000} \tag{8.39}$$

$$= \frac{LFN}{5250} \tag{8.40}$$

Brake Mean Effective Pressure, or Brake Mean Pressure

On high-speed engines and compressors, it is not possible to take indicator cards, but brake horsepower (bhp) readings can be expressed as equivalent brake mean pressure by combining results of Eq. (8.39) in Eq. (8.40), or

$$\text{Brake mean pressure, psi (kPa)} = \frac{\text{bhp}}{Lan} \times 33,000 \tag{8.41}$$

Mean Friction Pressures

Mean friction pressure measures the losses between cylinders and shaft.
   On engines,

Mean friction pressure, psi (kPa)

= indicated mean effective pressure − brake mean pressure (8.42)

On compressors and pumps,

Mean friction pressure, psi (kPa)

= brake mean pressure − indicated mean pressure (8.43)

Mechanical Efficiency

Mechanical efficiency is another way of expressing the losses between cylinder and shaft.

On engines,

$$\text{Mechanical eff} = \frac{\text{bhp}}{\text{indicated hp}} \times 100 \qquad (8.44)$$

On pumps and compressors,

$$\text{Mechanical eff} = \frac{\text{indicated hp}}{\text{bhp}} \times 100 \qquad (8.45)$$

HEAT ENGINE CYCLES—IDEAL

Carnot Cycle

The maximum thermal efficiency for the conversion of heat to work is specified by the Carnot cycle (Fig. 8.12). That efficiency is independent of the properties of the working substance and is defined as follows:

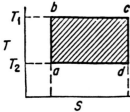

FIGURE 8.12   **FIGURE 8.13**

$$\text{Thermal eff} = \frac{\text{work done}}{\text{heat added}}$$

$$= \frac{Q_1 - Q_2}{Q_1} = \frac{T_1 - T_2}{T_1} \tag{8.46}$$

The mean effective pressure of a Carnot cycle depends upon the properties of the working substance. For a fixed gas (Fig. 8.13)

$$\text{Thermal eff} = 1 - \left(\frac{1}{R_v}\right)^{k-1} \tag{8.47}$$

Figure 8.14 shows the result of applying Eq. (8.47) to the Carnot cycle, using air as the working substance ($k = 1.4$). The work of the cycle, where the subscripts refer to Fig. 8.13, is

$$\Delta W_{\text{cycle}} = (T_2 - T_1) \frac{WR}{J} \ln \frac{v_c}{v_b} \tag{8.48}$$

and

$$\text{mep} = \frac{(T_2 - T_1)WR \ln (v_c/v_b)}{v_d - v_b} \tag{8.49}$$

For a wet vapor, like steam (Fig. 8.15)

$$W_{\text{cycle}} = \text{area } abcd \tag{8.50}$$

$$\text{mep} = \frac{W_{\text{cycle}}}{v_d - v_b} \tag{8.51}$$

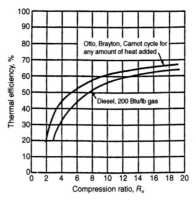

FIGURE 8.14

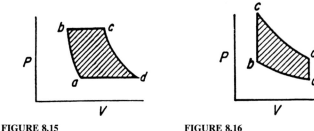

FIGURE 8.15   **FIGURE 8.16**

Otto Cycle

This is the basic cycle used in gasoline and other mixture engines (Fig. 8.16), and

$$\text{Thermal eff} = 1 - \left(\frac{1}{R_v}\right)^{k-1} \tag{8.52}$$

Values are shown graphically for air in Fig. 8.15.

$$\text{Work, ft} \cdot \text{lb} = JQ_{\text{added}} \times \text{thermal eff} \tag{8.53}$$

$$\text{mep, psi} = \frac{\text{work}}{144(v_a - v_b)} \tag{8.54}$$

where subscripts refer to Fig. 8.16.

Diesel Cycle

This is the basic cycle for injection-type internal combustion engines (Fig. 8.17), and

$$\text{Thermal eff} = 1 - \frac{1}{(v_a/v_b)^{k-1}} \frac{(v_c/v_d)^k - 1}{K(v_c/v_d - 1)} \tag{8.55}$$

Values for the air card standard are shown in Fig. 8.15.

$$\text{Work} = J[WC_p(T_c - T_b) - WC_v(T_d - T_a)] \tag{8.56}$$

$$\text{mep} = \frac{\text{work}}{v_a - v_b} \tag{8.57}$$

Heat recoverable from exhaust may be estimated by

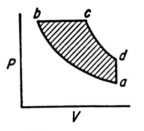

FIGURE 8.17

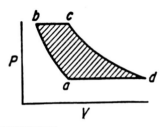

FIGURE 8.18

$$Q = \text{shp } C \frac{\Delta T}{4} \qquad (8.58)$$

Brayton Cycle

This is the cycle which prevails with gas-turbine power plants (Fig. 8.18), and

$$\text{Thermal eff} = 1 - \left(\frac{1}{R_v}\right)^{k-1} \qquad (8.59)$$

For the air card standard, the graphical data of Fig. 8.15 apply.

$$\text{Work} = WC_p J(T_c - T_b - T_d + T_a) \qquad (8.60)$$

$$\text{mep} = \frac{\text{work}}{v_d - v_b} \qquad (8.61)$$

Rankine Cycle

The ideal cycle involving vapors such as steam is evaluated only with the real physical properties of the fluid as given in steam tables and Mollier charts. From the general energy equation the work of the prime mover ΔW_{pm} is evaluated as shown in Figs. 8.19 and 8.20, in which

$$\Delta W_{pm} = h_1 - h_2 \qquad \text{Btu/lb (kJ/kg)} \qquad (8.62)$$

where h_1 = throttle enthalpy, Btu/lb, and h_2 = exhaust enthalpy (at same entropy as h_1), Btu/lb. The water rate or steam rate is

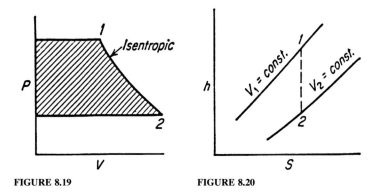

FIGURE 8.19   **FIGURE 8.20**

$$\text{Water rate} = \frac{3412.75}{\Delta W_{\text{pm}}} \qquad \text{lb/kWh} \qquad (8.63)$$

$$= \frac{2544.1}{\Delta W_{\text{pm}}} \qquad \text{lb/(hp} \cdot \text{h)} \qquad (8.64)$$

The feed pump work ΔW_{fp} required to deliver the water to the boiler (Fig. 8.21) is

$$\Delta W_{\text{fp}} = h_{c'} - h_b \qquad \text{Btu/lb (kJ/kg)} \qquad (8.65)$$

where h_b = enthalpy of liquid entering pump, Btu/lb (kJ/kg), and $h_{c'}$ = enthalpy of liquid leaving pump (same entropy as at h_b), Btu/lb (kJ/kg).

If water is noncompressible, then c' coincides with c in Fig. 8.21 and

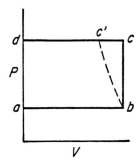

FIGURE 8.21

$$\text{Btu/lb} = \frac{\Delta p \times 144 \times \overline{v}}{778} = \frac{\text{head on pump}}{778} \qquad (8.66)$$

Net work of cycle $= \Delta W_{net} = \Delta W_{pm} - \Delta W_{fp}$ Btu/lb (kJ/kg)   (8.67)

Heat added to generate steam $= \Delta Q_{added}$

$$= h_1 - h_{2_{liq}} - \Delta W_{fp} \text{ (Btu/lb (kJ/kg))}$$
$$(8.68)$$

where $h_{2_{liq}}$ = enthalpy of saturated liquid at the exhaust of the prime mover.

$$\text{Thermal eff of Rankine cycle} = \frac{\Delta W_{net}}{\Delta Q_{added}}$$
$$= \frac{h_1 - h_2 - \Delta W_{fp}}{h_1 - h_{2_{liq}} - \Delta W_{fp}} \qquad (8.69)$$

For low-pressure cycles ΔW_{fp} becomes negligibly small so that

$$\text{Thermal eff of Rankine cycle} \cong \frac{h_1 - h_2}{h_1 - h_{2_{liq}}} \qquad (8.70)$$

$$\text{Heat rate of Rankine cycle, Btu/kWh supplied} = \frac{3412.75}{\text{thermal eff}} \qquad (8.71)$$

THROTTLING CALORIMETER

The throttling calorimeter, used for measuring the enthalpy of wet vapor, operates on the basis of constant-enthalpy expansion from conditions in high-pressure line to pressure in calorimeter chamber. Use the Mollier chart, found in most mechanical engineering handbooks, for graphical solution where $h_{\text{calorimeter chamber}} = h_{\text{high-pressure line}}$.

GAS TURBINES

A simple open-cycle gas turbine and its T-S diagrams are shown in Fig. 8.22. Figure 8.23 gives the work formulas for the compressor, combustor, and turbine in this simple open-cycle gas turbine. The P-V diagrams for each element in the cycle are also shown. Abbreviations and symbols used

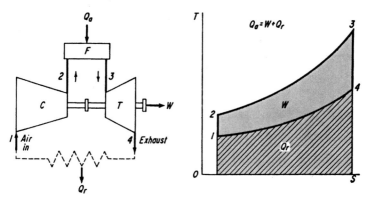

FIGURE 8.22

in Fig. 8.23 are given in Fig. 8.24. Alternative formulas for the simple open-cycle gas turbine are

$$T_2 = T_1 P_r^{(k-1)/k} \tag{8.72}$$

$$W_c = c_p(T_2 - T_1) \tag{8.73}$$

$$Q_a = c_p(T_3 - T_2) \tag{8.74}$$

$$T_4 = \frac{T_3}{P_r^{(k-1)/k}} \tag{8.75}$$

$$W_t = c_p(T_3 - T_4) \tag{8.76}$$

$$\text{Thermal eff} = \frac{W}{Q_a} = \frac{W_t - W_c}{Q_a} \tag{8.77}$$

$$\text{Thermal eff} = 1 - \frac{1}{P_r^{(k-1)/k}} \tag{8.78}$$

$$\text{Energy in exhaust} = c_p(T_4 - T_1) \qquad \text{Btu/lb air} \tag{8.79}$$

$$Q_r = Q_a - W \tag{8.80}$$

$$P_3 = P_1 P_r \tag{8.81}$$

Figure 8.25 shows a regenerative gas-turbine cycle. Formulas for its performance are

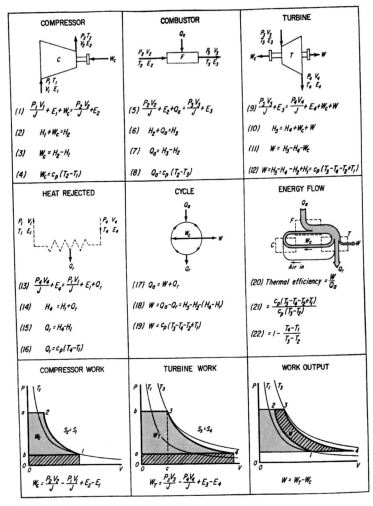

FIGURE 8.23

$$Q_t = c_p(T_5 - T_6) = c_p(T_3 - T_2) \qquad \text{Btu/lb air} \qquad (8.82)$$

$$Q_a = c_p(T_4 - T_3) \qquad\qquad\qquad\qquad (8.83)$$

$$W_t = c_p(T_4 - T_5) \qquad\qquad\qquad\qquad (8.84)$$

THERMODYNAMICS FORMULAS

a = acceleration, ft/s^2

A = area, ft^2

bdc = bottom dead center

bhphr = brake horsepower-hour

bmep = brake mean effective pressure

bsfc = brake specific fuel consumption

Btu = British thermal unit

c = percent compressor clearance

c = specific heat, Btu/lb

c_a = polytropic specific heat, Btu/lb

c_b = specific heat at constant pressure, Btu/lb

c_v = specific heat at constant volume, Btu/lb

C = carbon atom

C_d = nozzle coefficient of discharge

C_v = nozzle velocity coefficient

CO = carbon monoxide molecule

CO_2 = carbon dioxide molecule

COP = coefficient of performance

d = dimension, ft

D = gas density, lb/ft^3

e = cycle thermal efficiency

e_b = boiler efficiency

e_c = compressor efficiency

e_e = engine efficiency

e_n = nozzle efficiency

e_r = regenerator effectiveness

e_s = stage efficiency

e_t = thermal efficiency

e_v = compressor volumetric efficiency

e_D = diffuser efficiency

e_N = nozzle efficiency

E = internal energy, Btu/lb

E_k = kinetic energy, Btu or ft · lb/lb

E_p = potential energy, ft · lb/lb

F = force lb (= ma)

F = temperature, degrees Fahrenheit

g = acceleration of gravity = 32.2 ft/s^2

h_f = enthalpy of liquid, Btu/lb

h_{fe} = enthalpy of vaporization, Btu/lb

h_v = enthalpy of vapor Btu/lb

h-p = high pressure

H = enthalpy, Btu or ft · lb/lb

H = hydrogen atom

H_s = stagnation enthalpy, Btu/lb = $H_1 + E_{k1}$

H_2O = water-vapor molecule

HHV = higher heating value, Btu/lb

HR = heat rate, Btu/kwhr

Ihp = indicated horsepower

J = 778.26 ft · lb/Btu

in. Hg abs = inches of mercury absolute

k = ratio of specific heats = c_p/c_o

KE = kinetic energy, ft · lb or Btu/lb

kwhr = kilowatthour = 3412.75 Btu

$\log_e$ = logarithm to the base e

l-p = low pressure

L = piston stroke, ft

m = mass = w/g

m = bleed flow, lb per lb throttle steam

M = mass

M = Mach number

M = molecular weight

n = polytropic-process constant

N = total number of items

N = nitrogen atom

O = oxygen atom

psf = lb/ft^2 pressure

psfa = lb/ft^2 absolute

psfg = lb/ft^2 gauge

psi = lb/in^2

psia = lb/in^2 absolute

psig = lb/in^2 gauge

P = pressure, psi, psf, psia, psig psfa, psfg

FIGURE 8.24 Abbreviations and symbols used in Fig. 8.23.

P_c = nozzle critical pressure, psia, psfa

P_m = mean effective pressure, psi, psf

P_s = stagnation pressure, psia, psfa

P_r = pressure ratio

P_r = reduced pressure

Q = heat transferred, Btu/lb

Q_c = heat added to cycle Btu/lb

Q_r = heat rejected from cycle, Btu/lb

R = gas constant

R = absolute temperature, degree Rankine

R_u = universal gas constant = 1545

RF = reheat factor

$\overline{RH}$ = relative humidity, percent

s = distance or length, ft

s_f = entropy of liquid, Btu/lb · °F

s_{fs} = entropy increase of saturated liquid to vapor, Btu/lb · °F

s_v = entropy of vapor, Btu/lb · °F

S = entropy, Btu/lb · °F

S = sulfur atom

$\overline{SH}$ = specific humidity, lb vapor per lb dry gas

SR = steam rate, lb/kwhr

t = temperature, °F

t = time, s

tdc = top dead center

T = absolute temperature, °R

T_c = nozzle critical temperature, °R

T_s = stagnation temperature, °R

T_r = receiver temperature, °R

T_r = reduced temperature

T_s = source temperature

$\overline{TH}$ = total heat Btu per lb mixture

u = velocity, fps (ft/s)

v = velocity, fps

v_f = specific volume of saturated liquid, ft³/lb

v_{fs} = specific volume increase of saturated liquid to vapor, ft³/lb

v_s = specific volume of vapor, ft³/lb

V = specific volume, ft³/lb

V_c = clearance volume, ft³

V_d = displacement volume, ft³

V_M = mole volume, ft³

V_r = compression ratio = V_1/V_2

V_r = reduced volume

v_t = total volume, ft³

w = weight of a mass, lb

W = mechanical work, ft · lb or Btu/lb

W_e = engine work output, Btu or ft · lb/lb

W_f = flow work, ft · lb or Btu/lb

W_H = work in h-p cylinder ft · lb or Btu/lb

W_L = work in l-p cylinder, ft · lb or Btu/lb

W_o = lb oxygen per lb fuel

W_p = pump work input, Btu or ft · lb/lb

z = percent quality

y = percent moisture

Z = compressibility factor

FIGURE 8.24 (*Continued*) Abbreviations and symbols used in Fig. 8.23.

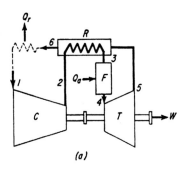

(a)

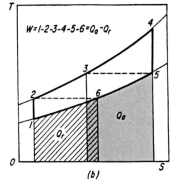

(b)

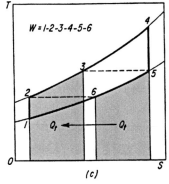

(c)

FIGURE 8.25

$$e_t = \frac{W}{Q_a} = \frac{W_t - W_c}{Q_a} \qquad (8.85)$$

$$e_t = 1 - \frac{T_1}{T_4} P_r^{(k-1)/k} \qquad (8.86)$$

$$Q_r = Q_a - W \qquad (8.87)$$

$$P_4 = P_1 P_r \qquad (8.88)$$

SECTION 9
ENERGY ENGINEERING FORMULAS

POWER PLANT PERFORMANCE FACTORS

Heat rate defined for the overall thermal performance as

$$\text{Heat rate, Btu/kWh} = \frac{\text{heat supplied in fuel for period, Btu}}{\text{energy output for period, kWh}}$$

$$(\text{Btu/kWh} \times 2.33 = \text{kJ/kWh}) \quad (9.1)$$

$$\text{Thermal eff, \%} = \frac{3412.75}{\text{heat rate}} \times 100 \quad (9.2)$$

$$\text{Capacity factor, \%} = \frac{\text{average load for period, kW}}{\text{rated capacity, kW}} \times 100 \quad (9.3)$$

$$\text{Load factor, \%} = \frac{\text{average load for period, kW}}{\text{peak load during period, kW}} \times 100 \quad (9.4)$$

BOILER PERFORMANCE

Nomenclature

h_{steam} = enthalpy leaving boiler unit (superheater outlet), Btu/lb (kJ/kg)

$h_{\text{feedwater}}$ = enthalpy of feedwater entering boiler unit (economizer inlet), Btu/lb (kJ/kg)

W_m = moisture content of fuel, lb/lb (kg/kg)

t_{fuel} = fuel temperature, °F (°C)

t_{fg} = flue-gas temperature, °F (C)

H_2 = lb hydrogen/lb fuel, from ultimate analysis (kg/kg)

W_{da} = weight of dry air supplied, lb/lb fuel (kg/kg)

W_w = weight water vapor/lb dry air (kg/kg)

t_a = ambient temperature or temperature of air entering air heater, °F (°C)

W_{dg} = weight dry flue gases, lb/lb fuel (kg/kg)

C = lb carbon/lb fuel, from ultimate analysis (kg/kg)

CO = CO in flue gas, dry volumetric basis, percent

CO_2 = CO_2 in flue gas, dry volumetric basis, percent

refuse = lb refuse/lb fuel, as burned in boiler furnace (kg/kg)

ash = lb ash/lb fuel, from ultimate analysis (kg/kg)

Heat Added to Steam

$$\Delta Q = h_{\text{steam}} - h_{\text{feedwater}} \qquad \text{Btu/lb (kJ/kg)} \tag{9.5}$$

With a resuperheater, the heat added as reheat must be included, and

$$h_{\text{reheat}} = h_{\text{leaving reheater}} - h_{\text{entering reheater}} \tag{9.6}$$

Boiler Rating and Steam Output

In small "packaged" boilers, the term *developed boiler horsepower* is used to measure the output of the boiler, or the heat added to the steam; it is defined as the evaporation of 34.5 (15.7 kg) water from and at 212°F (100°C). Thus,

Developed boiler hp = $34.5 \times 970.4 = 33,479$ Btu/h (9.8 kW)

The rated boiler horsepower is defined as

Rated boiler hp = 10 ft^2 (0.920 m^2) heating surface   (9.7)

Thus,   Percent rating = $\dfrac{\text{developed boiler hp}}{\text{rated boiler hp}} \times 100$ (9.8)

Factor of Evaporation (FE)

$$\text{FE} = \frac{\text{actual heat absorbed in converting water to steam}}{\text{latent heat of steam from and at 212°F (100°C)}} \tag{9.9}$$

$$= (h_{\text{steam}} - h_{\text{feedwater}})/970.4 \text{ (2262 kJ/kg)}$$

Evaporation

Actual evaporation AE, lb steam/lb fuel (kg/kg)

$$= \frac{\text{lb (kg) steam made during period}}{\text{lb (kg) fuel fired during period}} \qquad (9.10)$$

Equivalent evaporation EE, lb steam/lb fuel (kg/kg) from and at 212°F (100°C)

$$= \text{actual evaporation} \times \text{factor of evaporation AE} \times \text{FE} \qquad (9.11)$$

Boiler Efficiency

$$\text{Boiler eff} = \frac{\text{heat added to steam over period, Btu (kJ)}}{\text{heat supplied in fuel over period, Btu (kJ)}} \qquad (9.12)$$

The heat supplied in the fuel is the high or gross heating value on the *as-fired basis*.

Heat Balance and Losses

By the first law of thermodynamics it is possible to account for all the heat supplied in the fuel by adding all the losses to the heat supplied to the steam.

Loss due to moisture in fuel

$$= W_m(1090.7 - t_{\text{fuel}} + 0.455t_{\text{fg}}) \qquad \text{Btu/lb fuel†} \quad (9.13)$$

Loss due to hydrogen burning to water vapor instead of liquid

$$= 9 \times H_2(1090.7 - t_{\text{fuel}} + 0.455t_{\text{fg}}) \qquad \text{Btu/lb fuel} \quad (9.14)$$

where H_2 = lb hydrogen/lb fuel, from ultimate analysis (kg/kg).

Loss due to moisture in air

$$= W_{\text{aa}} \times W_w \times 0.47(t_{\text{fg}} - t_a) \qquad \text{Btu/lb fuel} \quad (9.15)$$

Loss due to dry stack gases $= 0.24 W_{\text{dg}}(t_{\text{fg}} - t_a) \qquad \text{Btu/lb fuel} \quad (9.16)$

---

† Btu/lb × 2.33 = kJ/kg.

STEAM BOILERS

Steam boiler efficiency is computed using these formulas (Fig. 9.1):

$$\frac{\text{Steam generator}}{\text{overall efficiency}} = \frac{\text{output, Btu/h}}{\text{input, Btu/h}} \qquad (9.17)$$

$$\text{Output, Btu/h} = S(h_g - h_{f1}) + S_r(h_{g3}/h_{g2}) + B(h_{f3} - h_{f1}) \qquad (9.18)$$

where S = steam flow, lb/h
  S_r = reheated steam flow, lb/h (if any)
  B = blowoff, lb/h

$$\text{Input, Btu/h} = FH \qquad (9.19)$$

where F = fuel input, lb/h (as fired), and H = fuel higher heating value, Btu/lb (as fired).

h_f = enthalpy of feedwater, Btu/lb
h_g = enthalpy of steam leaving the boiler, Btu/lb

$$\frac{\text{Economizer}}{\text{efficiency}} = \frac{\text{heat absorbed, Btu/h}}{\text{heat available, Btu/h}} \qquad (9.20)$$

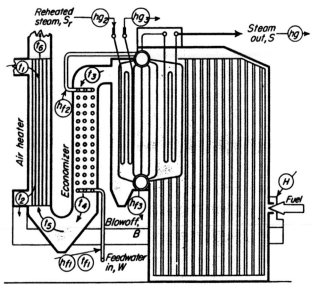

FIGURE 9.1 Measurement points for temperatures and enthalpies used in computing steam boiler efficiency.

$$\text{Heat absorbed, Btu/h} = W(h_{f2} - h_{f1}) \tag{9.21}$$

where W = feedwater flow, lb/h.

$$\text{Heat available, Btu/h} = H_g F \tag{9.22}$$

where H_g = heat available in flue gas, Btu/lb fuel
 = heat available in dry gas + heat available in flue-gas vapor, Btu/lb fuel

$$= (t_3 - t_{f1})0.24G + (t_3 - t_{f1})0.46 \left\{ M_f + 8.94H_2 \right.$$
$$\left. + M_a \left[G - C_b - N_2 - 7.94 \left(H_2 - \frac{O_2}{8} \right) \right] \right\} \tag{9.22a}$$

$$G = \frac{11CO_2 + 8O_2 + 7(N_2 + CO)}{3(CO_2 + CO)} \times \left(C_b + \frac{S}{2.67} \right) + \frac{S}{1.60} \tag{9.23}$$

where M_f = lb moisture/lb fuel burned
 M_a = lb moisture/lb dry air to furnace
 C_b = lb carbon burned/lb fuel burned = $C - Rc_r$
 C_r = lb combustible/lb refuse

Conversion factors for the formulas above are as follows: Btu/h × 0.293 = W; lb/h × 0.454 = kg/h; Btu/lb × 2.33 = kJ/kg; lb/lb = kg/kg. Figure 9.1 shows where temperatures and enthalpies are measured for computing steam boiler efficiency.

$$\frac{\text{Air heater}}{\text{efficiency}} = \frac{\text{heat absorbed, Btu/lb fuel}}{\text{heat available, Btu/lb fuel}} \tag{9.24}$$

$$\text{Heat absorbed, Btu/lb fuel} = A_h(t_2 - t_1)(0.24 + 0.46M_a) \tag{9.25}$$

where A_h = airflow through heater, lb/lb fuel = $A - A_m$
 A = total air to furnace, lb/lb fuel

$$= G - C_b - N_2 - 7.94 \left(H_2 - \frac{O_2}{8} \right)$$

 G = similar to economizer but based on gas at furnace exit
 A_m = external air supplied by mill fan or other source, lb/lb fuel

Heat available, Btu/lb fuel = $(t_5 - t_1)0.24G + (t_5 - t_1)0.46$
$$(M_f + 8.94H_2 + M_a A) \tag{9.26}$$

where G is the same as for economizer, above, and A is as defined, both based on the characteristics of the gas entering the air heater.

FUELS AND COMBUSTION

Fuel Heating Value

The heating value of solid fuels, such as coal, coke, and bagasse, can be found from

$$Q = 14{,}500C + 62{,}000 \left(H - \frac{O}{8} \right) + 4000S \qquad (9.27)$$

where Q = heating value, Btu/lb as fired; C = percent carbon (volatile and fixed, also called *total carbon*) in the fuel, expressed as a decimal; H = percent hydrogen; O = percent oxygen, and S = percent sulfur in the fuel, all expressed in decimal form. To convert to kilojoules, multiply the Btu value by 1.055.

The heating value of liquid fuel—oil—is given by

$$Q = 13{,}500C + 60{,}890H \qquad \text{Btu} \qquad (9.28)$$

where the symbols are the same as given above.

When the Baumé reading for a liquid fuel is known, the heating value is given by

$$Q = 18{,}650 + 40(\text{Baumé reading} - 10) \qquad (9.29)$$

where Bé is the Baumé reading of the liquid fuel.

Air Requirements The air requirements for various fuels—solid, liquid, and gaseous—are

$$\text{Air required, lb/lb fuel} = \frac{\text{higher heating value of fuel, Btu/lb}}{1300}$$

$$(\text{lb/lb} \times 0.454 = \text{kg/kg}) \qquad (9.30)$$

Products of Combustion

Flue gas analyses are made to determine the effectiveness of combustion operations and are ordinarily given on the dry, volumetric basis. If the nitrogen content of the fuel is small, then the excess air can be computed from

$$\text{Excess air} = \frac{3.78(O_2 - CO/2)}{N_2 - 3.78(O_2 - CO/2) \times 100} \qquad \text{percent} \qquad (9.31)$$

where O_2, N_2, and CO are percentages by volume obtained from the flue gas analysis. Some customary excess-air values are given in data tables in engineering handbooks.

STEAM TURBINES

Four important measures of steam turbine performance are the steam rate, heat rate, thermal efficiency, and engine efficiency. Figure 9.2 shows the basic cycles on which turbine performance is computed. Formulas used to compute these measures are as follows:

Heat Rate

For all types of turbines

$$\text{Steam rate, lb/kWh} = \frac{\text{steam flow, lb/h } (W_1)}{\text{generator or shaft output, kWh } (P)} \quad (9.32)$$

For backpressure turbines

$$\text{Heat rate, Btu/kWh} = \frac{W_1(h_1 - h')}{P} \quad (9.33)$$

where h_1 = steam enthalpy entering turbine, Btu/lb (see diagram) and when steam exhausts to process,

$$h' = \text{actual exhaust enthalpy, Btu/lb}$$

and when steam exhausts to waste,

$$h' = \text{saturated-water enthalpy at exhaust pressure,}$$
$$\text{Btu/lb (see diagram)}$$

$$\text{Engine efficiency} = \frac{3413P}{(W_1 - W_g)(h_1 - h_s) + W_g(h_1 - h_{sg})} \quad (9.34)$$

where W_g = steam leaving turbine system from glands and leaks, lb/h
h_s = exhaust-steam enthalpy at entropy of initial steam, Btu/lb
h_{sg} = leakoff steam enthalpy at entropy of initial steam, Btu/lb

For straight-condensity turbines,

$$\text{Heat rate, Btu/kWh} = \frac{(W_1 - W_g)(h_1 - h_f) + W_g(h_1 - h_{fg})}{P} \quad (9.35)$$

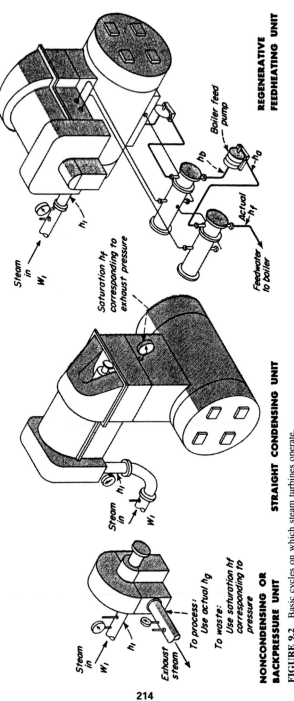

REGENERATIVE FEEDHEATING UNIT

Steam in w_i h_1

Boiler feed pump h_b h_a

Saturation h_f corresponding to exhaust pressure

Actual h_f

Feedwater to boiler

STRAIGHT CONDENSING UNIT

Steam in w_i h_1

NONCONDENSING OR BACKPRESSURE UNIT

Steam in w_i h_1

Exhaust steam

To process: Use actual h_g

To waste: Use saturation h_f corresponding to pressure

FIGURE 9.2 Basic cycles on which steam turbines operate.

214

where h_f = saturated-water enthalpy at exhaust pressure, Btu/lb (see diagram)

h_{fg} = vaporization enthalpy of leakoff steam at discharge press, Btu/lb

$$\text{Engine efficiency} = \frac{3413P}{(W_1 - W_g)(h_1 - h_s + W_g(h_1 - h_{sg})} \quad (9.36)$$

For regenerative turbines,

$$\text{Heat rate, Btu/kWh} = \frac{W_1(h_1 - h_f) + P_1(h_a - h_b)}{P} \quad (9.37)$$

where h_f = enthalpy of feedwater leaving last heater, Btu/lb

h_a = enthalpy of feedwater leaving boiler feed pump, Btu/lb

h_b = enthalpy of feedwater entering boiler feed pump, Btu/lb (see diagram)

$$\text{Engine efficiency} = \frac{3413P}{\begin{array}{c} W_{b1}(h_1 - h_{sb1}) + W_{b2}(h_1 - h_{sb2}) \\ + \cdots + W_{bn}(h_1 - h_{sbn}) \\ + W_g(h_1 - h_{sg}) \end{array}} \quad (9.38)$$

where $W_{b1}; W_{b2}, W_{bn}$ = bleed steam flows, lb/h

$h_{sb1}, h_{sb2}, h_{sbn}$ = enthalpies of bleed steam at initial steam entropy, Btu/lb

W_e = exhaust steam flow, lb/h

h_{se} = exhaust steam enthalpy at initial steam entropy, Btu/lb

For all types of turbines,

$$\text{Thermal efficiency} = \frac{3413}{\text{Heat rate, Btu per kWh}} \quad (9.39)$$

The following conversion factors can be used for the formulas above: lb/kWh $\times$ 0.126 = kg/MJ; lb/h $\times$ 0.454 = kg/h; Btu/kWh $\times$ 0.95 = kJ/kWh; Btu/lb $\times$ 2.33 = kJ/kg.

Steam Rate for Reheat-Regenerative Cycle

1. Assemble the key enthalpies, entropies, and pressures for the cycle (Fig. 9.3).
2. Compute the percent steam extracted for the feedwater heater. The steam extracted for the feedwater heater $x = (H_{fx} - H_f)(H_x - H_f)$.
3. Find the turbine steam rate. For the Rankine-cycle steam rate, $w_s = 3413/(H_1 - H_c)$.

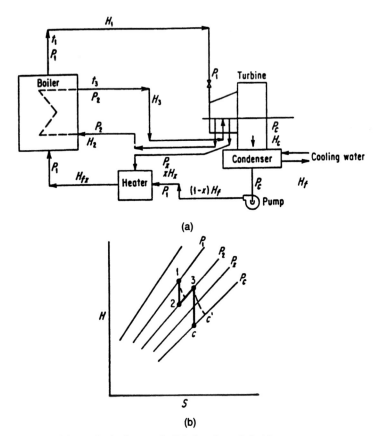

FIGURE 9.3   (a) Cycle diagram. (b) H-S chart for cycle in (a).

4. Calculate the turbine thermal efficiency. The thermal efficiencies $E_t = [(H_1 - H_2) + x(H_3 - H_x) + (1 - x)(H_3 - H_c)]/(H_3 - H_2 + H_1 - H_{fx})$.
5. Determine the condition of the exhaust. The engine efficiency of the turbine alone equals actual turbine combined efficiency divided by actual generator efficiency.

Steam Turbogenerator Efficiency and Steam Rate

The combined thermal efficiency (CTE) = $(3413/w_r)[1/(h_1 - h_2)]$, where w_r = combined steam rate, lb/kWh (kg/kWh); h_1 = enthalpy of steam at throttle pressure and temperature, Btu/lb (kJ/kg); and h_2 = enthalpy of steam at the turbine backpressure, Btu/lb (kJ/kg), using the steam tables and Mollier chart.

The combined engine efficiency (CEE) = w_i/w_e = (weight of steam used by ideal engine, lb/kWh)/(weight of steam used by actual engine, lb/kWh). The weights of steam used may also be expressed as btu/lb (kJ/kg). Thus, for the ideal engine, the value is 3413 Btu/lb (7952.3 kJ/kg). For the actual turbine, $h_1 - h_{2x}$ is used, h_{2x} is the enthalpy of the wet steam at exhaust conditions; h_1 is as before.

To find the CEE, we first must obtain the ideal steam rate $w_i = 3413/(h_1 - h_{2x})$. Use this approach to analyze the efficiency of any turbogenerator used in central-station, industrial, marine, and other plants.

Turbogenerator Reheat-Regenerative Cycle: Alternatives Analysis

1. Using the steam tables and Mollier chart, list the pertinent steam conditions (Fig. 9.4). With the subscript 1 for throttle conditions, list the key values for the cycle thus: P_1, t_1, h_1, S_1, H_2, H_3, H_4, H_5, and H_6.

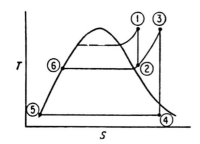

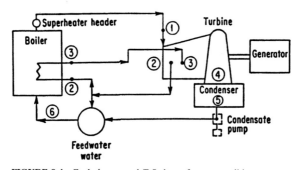

FIGURE 9.4   Cycle layout and *T-S* chart of steam conditions.

2. Determine the percentage of throttle steam bled for feedwater heating. Set up the ratio for the feedwater heater of (heat added in the feedwater heater)/(heat supplied to the heater)(100).

3. Find the heat converted to work per pound (kilogram) of throttle steam. The heat converted to work is the enthalpy difference between the throttle steam and the bleed steam at point 2 plus the enthalpy difference between points 3 and 4 times the percentage of throttle flow between these points. In equation form, heat converted to work = $H_1 - H_2 + (1.00 - p)(H_3 - H_4)$ where p = percentage of throttle steam bled for feedwater heating, in decimal form.

4. Calculate the heat supplied per pound (kilogram) of throttle steam. The heat supplied per pound (kilogram) of throttle steam = $H_1 - H_6 + H_3 - H_2$.

5. Compute the ideal thermal efficiency. Use this relation: ideal thermal efficiency = (heat converted to work)/(heat supplied).

Power Plant Performance Based on Test Data

1. Determine the steam properties at key points in the cycle. Using a Mollier chart and the steam tables, plot the cycle as in Fig. 9.5. The percent throttle steam bled is found from $100 \times (H_5 - H_4)/(H_2 - H_4)$.

2. Find the amount of heat converted to work. Use this relation: heat converted to work $h_w = H_1 - H_2 + (1 - m_2)(H_2 - H_7)$, where m_2 = percent throttle steam bled and H_7 = enthalpy of exhaust steam in the condenser.

3. Compute the ideal steam rate. Use this relation: ideal steam rate $r = (3413 \text{ Btu/kWh})/h_w$.

$$\text{Cycle efficiency } C_e = \frac{\text{heat converted into work}}{\text{heat supplied}}$$

4. Determine the combined steam rate. The combined steam rate for the actual unit is R_c = lb steam consumed/kWh generated.

5. Find the combined thermal efficiency of the actual unit. The combined thermal efficiency TE_c = 3413/heat supplied.

6. Compute the combined engine efficiency. The combined engine efficiency = TE_c/C_c.

HYDROELECTRIC PLANTS AND HYDRAULIC TURBINES

Nomenclature

Q = flow, ft³/s (m³/s)
H = head on site, ft (m)

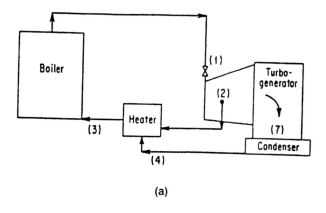

(a)

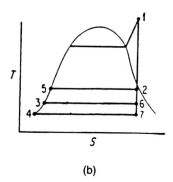

(b)

FIGURE 9.5 Plot of steam cycle.

σ = cavitation coefficient

H_b = barometer head, ft (m)

H_v = vapor pressure, ft (m)

H_s = static suction head, measured positively from tail race to runner-blade periphery, ft (m)

H_e = effective head on unit, ft (m)

Water Horsepower

The theoretical power output for a hydro installation is

$$\text{Water hp} = \frac{QH}{8.8} \qquad (9.40)$$

or
$$\text{Water kW} = \frac{QH}{11.8} \qquad (9.41)$$

Specific Speed

Specific speed in hydraulic turbine practice is defined differently than for pumps. Here the head on the site and the power output are so related that specific speed is defined as that at which a unit of suitable diameter of an homologous series would run in order to deliver 1 hp (0.754 kW) under 1-ft (0.3048-m) head, or

$$N_s = \text{specific speed} = \frac{\text{rpm} \times \text{shp}^{0.5}}{\text{head}^{1.25}} \qquad (9.42)$$

Cavitation

A water wheel must be set with reference to tail-water level at a height that avoids cavitation. A unit must not be designed to operate at lower values of the cavitation coefficient, as defined by

$$\sigma = H_b - H_v - \frac{H_s}{H_e} \qquad (9.43)$$

SURFACE CONDENSERS FOR STEAM TURBINES

The cooling-water flow for surface condensers (Fig. 9.6) is given by

$$G = \frac{950S}{500(t_2 - t_1)} = \frac{1.9S}{t_2 - t_1} \qquad (9.44)$$

where G = cooling-water flow, gal/min
S = steam condensed, lb/h
t_2 = outlet water temperature, °F
t_1 = inlet water temperature, °F

(*Note:* 950 Btu is assumed to be removed from one lb of steam.) The condenser tube surface area is

$$A = \frac{kL}{V} G \qquad (9.45)$$

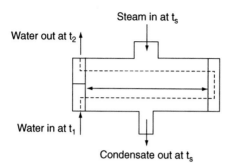

FIGURE 9.6 Surface condenser temperatures used in performance calculations.

where A = condenser surface area, ft²
k = constant given in engineering handbooks
L = tube length per pass, ft
V = water velocity, ft/s

The outlet water temperature from the condenser is

$$t_2 = t_N - \frac{t_N - t_1}{e^x} \qquad (9.46)$$

where $x = (k/V)(U/500)$
$e^x = 2.7183^x$ given in engineering handbooks
t_N = saturated steam temperature corresponding to absolute pressure in condenser shell, °F
U = heat-transfer rate, Btu/(ft² · h · °F) LMTD

BOILER HEAT BALANCE

To show the distribution of heat in 1 lb of fuel as fired, a boiler heat balance is usually constructed. The items to account for are

1. Heat absorbed by boiler [Eq. (9.47)]
2. Heat lost in dry exit gases [Eq. (9.48)]
3. Evaporation of moisture formed by burning hydrogen in fuel [Eq. (9.49)]
4. Evaporation of surface moisture in fuel [Eq. (9.50)]
5. Loss due to incomplete combustion [Eq. (9.51)]
6. Loss due to unconsumed carbon in the ash [Eq. (9.52)]
7. Heat lost owing to heating moisture in the air [Eq. (9.53)]

8. Radiation and unaccounted for losses

These items may be computed as follows:

$$(1) \quad h_1 = W(H - h_f) \tag{9.47}$$

where W = lb water actually evaporated/lb fuel fired
$\quad H$ = heat in1 lb steam at exit condition, i.e., at superheater outlet if superheated
$\quad h_f$ = heat in 1 lb feedwater to boiler

$$(2) \quad h_2 = W_g(T_g - t_r)c_p \tag{9.48}$$

where W_g = lb dry flue gas (from combustion diagram)/lb fuel fired
$\quad c_p$ = specific heat of gases (usually assumed = 0.24)
$\quad T_g$ = temperature exit flue gas
$\quad t_r$ = temperature combustion air to furnace

$$(3) \quad h_s = 9h[212 - t_c + 970.3 + 0.46(T_g - 212)] \tag{9.49}$$

where h = hydrogen as a fraction per pound fuel fired
$\quad 9h$ = amount water formed
$\quad t_c$ = temperature of fuel as fired

$$(4) \quad h_4 = w[212 - t_c + 970.3 + 0.46(T_g - 212)] \tag{9.50}$$

where w = lb surface moisture/lb fuel fired. Latest results should be considered; h_c = 14,150. Also 3960 Btu/lb is lost by all C converted to CO.

$$(5) \quad h_5 = \frac{CO}{CO + CO_2} (14,150 - 3960)C_b \tag{9.51}$$

where CO = percentage CO as from flue-gas analysis
$\quad CO_2$ = percentage CO_2 as from flue-gas analysis
$\quad C_b$ = weight carbon in 1 lb of fuel actually burned
Total carbon = $C = C_b + C_a$

$$(6) \quad h_6 = 14,150 \times \frac{W_a}{W} \times C_a \tag{9.52}$$

where W_a = weight ash collected per unit period
$\quad W$ = weight fuel fired per unit period
$\quad C_a$ = percentage combustible in ash (usually assumed to be carbon)

$$(7) \quad h_7 = M0.46(T_g - t_r) \tag{9.53}$$

where M = actual weight moisture per pound dry air at existing dry- and wet-bulb conditions.
(8) Difference between summation of all the preceding—(1) to (7)—and the heat value of 1 lb of fuel as fired.

PRESSURE DROP IN STRAIGHT DUCTS FOR BOILER DUCTS

Buffalo Forge Company proposed this formula, which is satisfactory for ranges of Reynolds numbers from 25,000 to 5,000,000, covering practically the whole range normally encountered in power plant work.

$$\Delta p = \frac{1.64 F L_\mu^{0.16} \rho^{0.84}}{d^{1.24}} \left(\frac{V}{1000}\right)^{1.84}$$

$$= \frac{0.03 F L}{d^{1.24}} \left(\frac{V}{1000}\right)^{1.84} \qquad (9.54)$$

for air at 70°F (21.1°C)

and 29.92 in (759.9 mm) barometer

where Δp = pressure drop, in of water

$$F = \begin{cases} 0.80 & \text{smooth tubing and glass} \\ 1.00 & \text{iron ducts and average steel pipes} \\ 1.20 & \text{brick, rough concrete, and heavily riveted piping} \end{cases}$$

L = length of pipe, ft (m)

μ = viscosity of air or gas, lb/(ft·s) [kg/(m·s)]

ρ = gas or air density, lb/ft^3 (kg/m^3)

d = diameter of duct, in (mm)

$$= \frac{2 \times \text{length} \times \text{width}}{\text{length} + \text{width}} \quad \text{rectangular ducts} \qquad (9.55)$$

V = velocity, ft/min (m/min)

The expression above may be used to solve for friction losses in straight runs of ductwork and for the friction in stacks.

U-TUBES, MANOMETERS, AND DRAFT GAUGES

U-Tube (Fig. 9.7a and b)

When the interface between the mercury and the fluid for which the pressure is wanted is K ft (m) below the point of attachment A and gives a reading of H_m ft (m),

$$p_A = H_m \overline{W}_m - K \overline{W}_A \qquad (9.56)$$

At A

$$H_A = \frac{H_m \overline{W}_m}{\overline{W}_A - K} \qquad (9.57)$$

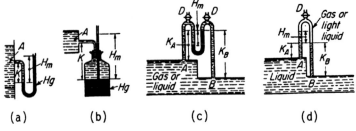

FIGURE 9.7 Manometers of several types.

where $\overline{W}_A$ = weight density of fluid at A, lb/ft^3 (kg/m^3)
$\overline{W}_m$ = density manometric fluid, lb/ft^3 (kg/m^3)
p_A = gauge pressure at A, lb/ft^2 (kPa)

Differential U-Tube

Figure 9.7c shows the difference between taps A and B to be

$$p_A - p_B = H_m(\overline{W}_m - \overline{W}_A) + K_A\overline{W}_A - K_B\overline{W}_B \qquad (9.58)$$

where K_A, K_B = vertical distances of upper mercury surface above A and B, ft (m)
$\overline{W}_A$, $\overline{W}_B$ = weight densities of fluid at A and B, lb/ft^3 (kg/m^3)

If the differential is that caused by an orifice or other device measuring the flow of a liquid, the orifice differential

$$\Delta H = p_2v_1 - p_2v_2 + Z_1 - Z_2 = H_m\left(\frac{\overline{W}_m}{\overline{W}_A} - 1\right) \qquad (9.59)$$

For gases, except at very high pressures, $\overline{W}_A$ and $\overline{W}_B$ are so small compared with $\overline{W}_m$ that Eq. (9.58) reduces to

$$p_A - p_B = H_m\overline{W}_m \qquad (9.60)$$

Inverted Differential U-Tube (Fig. 9.7d)

$$p_A - p_B = H_m(\overline{W}_A - \overline{W}_m) + K_A\overline{W}_A - K_B\overline{W}_B \qquad (9.61)$$

If the gauge is indicating the orifice differential of a head meter operating with a liquid,

$$\Delta H = H_m\left(1 - \frac{\overline{W}_m}{\overline{W}_A}\right) \qquad (9.62)$$

Closed U-Tubes

These measure directly the absolute pressure p of a fluid (Fig. 9.8a).

$$p = H_m \overline{W}_m \qquad (9.63)$$

where $\overline{W}_m$ = lbm/ft^3 (kgm/m^3) = weight density of manometric fluid
H_m = ft (m) manometric fluid

For liquids and gases under very high pressures, the quantity $K\overline{W}_0$ should be subtracted from Eq. (9.63).

Multiplying Gauges

Inclined U-Tube (Fig. 9.8c).   If the reading is R ft (m), the formula $H_m = (R - R_0) \sin \theta$ is substituted in Eq. (9.58), where R_0 = the zero reading.

The Draft Gauge.   Formulas are applied as with an inclined U-tube above (Fig. 9.8b).

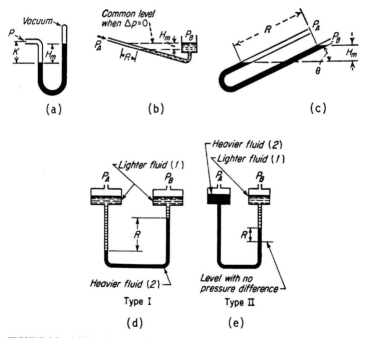

FIGURE 9.8   Additional types of manometers.

***Two-Fluid U-Tubes* (*Fig. 9.8d and e*).**   For type I:

$$p_A - p_B = R - R_0 \left(\overline{W}_2 - \overline{W}_1 + \frac{a}{A} \overline{W}_1 \right) \tag{9.64}$$

For type II:

$$p_A - p_B = R \left[\overline{W}_2 - \overline{W}_1 + \frac{a}{A} (\overline{W}_2 + \overline{W}_1) \right] \tag{9.65}$$

where A = cross-sectional area of each reservoir, ft^2 (m^2)
a = cross-sectional area of tube forming the U, ft^2 (m^2)

SECTION 10
FORMULAS FOR FLUIDS ENGINEERING

To simplify using the formulas in this section, Table 10.1 presents symbols, nomenclature, and U.S. Customary System (USCS) and Système International (SI) units found in each expression.

CAPILLARY ACTION

Capillarity is due to both the cohesive forces between liquid molecules and the adhesive forces of liquid molecules. It shows up as the difference in liquid surface elevations between the inside and outside of a small tube that has one end submerged in the liquid (Fig. 10.1).

Capillarity is commonly expressed as the height of this rise. In equation form,

$$h = \frac{2\sigma \cos \theta}{(w_1 - w_2)r} \qquad (10.1)$$

where h = capillary rise, ft (m)
  σ = surface tension, lb/ft (N/m)
  w_1, w_2 = specific weights of fluids below and above meniscus, respectively, lb/ft (N/m)
  θ = angle of contact
  r = radius of capillary tube, ft (m)

Capillarity, like surface tension, decreases with increasing temperature. Its temperature variation, however, is small and insignificant in most problems.

VISCOSITY

Viscosity μ of a fluid, also called the *coefficient of viscosity, absolute viscosity,* or *dynamic viscosity,* is a measure of the fluid's resistance to flow.

TABLE 10.1 Symbols, Terminology, Dimensions, and Units Used in Water Engineering

Symbol	Terminology	Dimensions	USCS units	SI units
A	Area	L^2	ft^2	mm^2
C	Chezy roughness coefficient	$L^{1/2}/T$	ft^5/s	$m^{0.5}/s$
C_1	Hazen-Williams roughness coefficient	$L^{0.37}/T$	$ft^{0.37}/s$	$m^{0.37}/s$
d	Depth	L	ft	m
d_c	Critical depth	L	ft	m
D	Diameter	L	ft	m
E	Modulus of elasticity	F/L^2	lb/in^2 (psi)	MPa
F	Force	F	lb	N
g	Acceleration due to gravity	L/T^2	ft/s^2	m/s^2
H	Total head, head on weir	L	ft	m
h	Head or height	L	ft	m
h_f	Head loss due to friction	L	ft	m
L	Length	L	ft	m
M	Mass	FT^2/L	$lb \cdot s^2/ft$	$N \cdot s^2/m$
n	Manning's roughness coefficient	$T/L^{1/3}$	$s/ft^{1/3}$	$s/m^{1/3}$
P	Perimeter, weir height	L	ft	m
P	Force due to pressure	F	lb	N
p	Pressure	F/L^2	lb/ft^2	MPa
Q	Flow rate	L^3/T	ft^3/s	m^3/s
q	Unit flow rate	$L^3/T \cdot L$	$ft^3/(s \cdot ft)$	$m^3/(s \cdot m)$
r	Radius	L	ft	m
R	Hydraulic radius	L	ft	m
T	Time	T	s	s
t	Time, thickness	T, L	s, ft	s, m
V	Velocity	L/T	ft/s	m/s
W	Weight	F	lb	kg
w	Specific weight	F/L^3	lb/ft^3	kg/m^3
y	Depth in open channel, distance from solid boundary	L	ft	m
Z	Height above datum	L	ft	m
ϵ	Size of roughness	L	ft	m
μ	Viscosity	FT/L^2	$lb \cdot s/ft$	$kg \cdot s/m$
ν	Kinematic viscosity	L^2/T	ft^2/s	m^2/s
ρ	Density	FT^2/L^4	$lb \cdot s^2/ft^4$	$kg \cdot s^2/m^4$
σ	Surface tension	F/L	lb/ft	kg/m
τ	Shear stress	F/L^2	lb/in^2	MPa

It is expressed as the ratio of the tangential shearing stresses between flow layers to the rate of change of velocity with depth:

$$\mu = \frac{\tau}{dV/dy} \qquad (10.2)$$

where τ = shearing stress, lb/ft^2 (N/m^2)
$\quad V$ = velocity, ft/s (m/s)
$\quad y$ = depth, ft (m)

TABLE 10.1   Symbols, Terminology, Dimensions, and Units Used in Water Engineering (*Continued*)

Symbol	Terminology	Dimensions	USCS units	SI units
	Symbols for dimensionless quantities			

Symbol	Quantity
C	Weir coefficient, coefficient of discharge
C_c	Coefficient of contraction
C_v	Coefficient of velocity
F	Froude number
f	Darcy-Weisbach friction factor
K	Head-loss coefficient
R	Reynolds number
S	Friction slope—slope of energy grade line
S_c	Critical slope
η	Efficiency
sp gr	Specific gravity

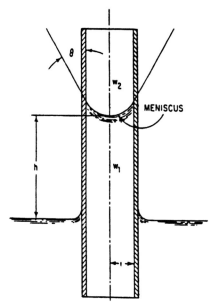

FIGURE 10.1   Capillary action raises water in a small-diameter tube. Meniscus, or liquid surface, is concave upward.

Viscosity decreases as temperature increases but may be assumed independent of changes in pressure for the majority of engineering problems. Water at 70°F (21.1°C) has a viscosity of 0.00002050 lb · s/ft^2 (0.00098 N · s/m^2).

Kinematic viscosity ν is defined as viscosity μ divided by density ρ. It is so named because its units, ft^2/s (m^2/s), are a combination of the kinematic units of length and time. Water at 70°F (21.1°C) has a kinematic viscosity of 0.00001059 ft^2/s (0.000001 m^2/s).

In hydraulics, viscosity is most frequently encountered in the calculation of the Reynolds number to determine whether laminar, transitional, or completely turbulent flow exists.

FUNDAMENTALS OF FLUID FLOW

For fluid energy, the law of conservation of energy is represented by the *Bernoulli equation:*

$$Z_1 + \frac{p_1}{w} + \frac{V_1^2}{2g} = Z_2 + \frac{p_2}{w} + \frac{V_2^2}{2g} \qquad (10.3)$$

where Z_1 = elevation, ft (m), at any point 1 of flowing fluid above an arbitrary datum

Z_2 = elevation, ft (m), at downstream point in fluid above same datum

p_1 = pressure at point 1, lb/ft^2 (kPa)

p_2 = pressure at point 2, lb/ft^2 (kPa)

w = specific weight of fluid, lb/ft^3 (kg/m^3)

V_1 = velocity of fluid at point 1, ft/s (m/s)

V_2 = velocity of fluid at point 2, ft/s (m/s)

g = acceleration due to gravity = 32.2 ft/s^2 (9.81 m/s^2)

The left side of the equation sums the total energy per unit weight of fluid at point 1; and the right side, the total energy per unit weight at point 2. Equation (10.3) applies only to an ideal fluid. Its practical use requires a term to account for the decrease in total head, ft (m), through friction. This term h_f, when added to the downstream side, yields the form of the Bernoulli equation most frequently used:

$$Z_1 + \frac{p_1}{w} + \frac{V_1^2}{2g} = Z_2 + \frac{p_2}{w} + \frac{V_2^2}{2g} + h_f \qquad (10.4)$$

The energy contained in an elemental volume of fluid thus is a function of its elevation, velocity, and pressure (Fig. 10.2). The energy due to elevation is the potential energy and equals WZ_a, where W is the weight, lb (kg), of the fluid in the elemental volume and Z_a is its elevation, ft (m), above some arbitrary datum. The energy due to velocity is the kinetic energy. It equals $WV_a^2/2g$, where V_a is the velocity, ft/s (m/s). The pressure

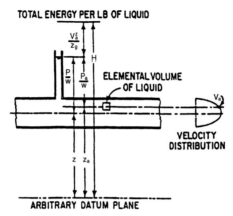

FIGURE 10.2 Energy in a liquid depends on elevation, velocity, and pressure.

energy equals Wp_a/w, where p_a is the pressure, lb/ft^2 (kg/kPa), and w is the specific weight of the fluid, lb/ft^3 (kg/m^3). The total energy in the elemental volume of fluid is

$$E = WZ_a + \frac{Wp_a}{w} + \frac{WV_a^2}{2g} \qquad (10.5)$$

Dividing both sides of Eq. (10.5) by W yields the energy per unit weight of flowing fluid, or the *total head*, ft (m):

$$H = Z_a + \frac{p_a}{w} + \frac{V_a^2}{2g} \qquad (10.6)$$

Here p_a/w is called the *pressure head* and $V_a^2/2g$, the *velocity head*.

As indicated in Fig. 10.2, $Z + p/w$ is constant for any point in a cross section and normal to the flow through a pipe or channel. Kinetic energy at the section, however, varies with velocity. Usually, $Z + p/w$ at the midpoint and the average velocity at a section are assumed when the Bernoulli equation is applied to flow across the section or when total head is to be determined. *Average velocity,* ft/s (m/s), is Q/A, where Q is the quantity of flow, ft^3/s (m^3/s), across the area of section A, ft^2 (m^2).

SIMILITUDE FOR PHYSICAL MODELS

A physical model is a system whose operation can be used to predict the characteristics of a similar system, or prototype, usually more complex or built to a much larger scale.

The ratios of the forces of gravity, viscosity, and surface tension to the force of inertia are designated, the Froude number, Reynolds number, and Weber number, respectively. Equating the Froude number of the model and the Froude number of the prototype ensures that the gravitational and inertial forces are in the same proportion. Similarly, equating the Reynolds numbers of the model and prototype ensures that the viscous and inertial forces are in the same proportion. Equating the Weber numbers ensures proportionality of surface tension and inertial forces.

The *Froude number* is

$$\mathbf{F} = \frac{V}{\sqrt{Lg}} \tag{10.7}$$

where $\mathbf{F}$ = Froude number (dimensionless)
V = velocity of fluid, ft/s (m/s)
L = linear dimension (characteristic, such as depth or diameter), ft (m)
g = acceleration due to gravity = 32.2 ft/s² (9.81 m/s²)

For hydraulic structures, such as spillways and weirs, where there is a rapidly changing water-surface profile, the two predominant forces are inertia and gravity. Therefore, the Froude numbers of the model and prototype are equated:

$$\mathbf{F}_m = \mathbf{F}_p \qquad \frac{V_m}{\sqrt{L_m G}} = \frac{V_p}{\sqrt{L_p g}} \tag{10.8}$$

where subscript m applied to the model and p to the prototype.

The *Reynolds number* is

$$\mathbf{R} = \frac{VL}{\nu} \tag{10.9}$$

Now $\mathbf{R}$ is dimensionless, and ν is the kinematic viscosity of fluid, ft²/s (m²/s). The Reynolds numbers of model and prototype are equated when the viscous and inertial forces are predominant. Viscous forces are usually predominant when flow occurs in a closed system, such as pipe flow where there is no free surface. The following relations are obtained by equating Reynolds numbers of the model and prototype:

$$\frac{V_m L_m}{\nu_m} = \frac{V_p L_p}{\nu_p} \qquad V_r = \frac{\nu_r}{L_r} \tag{10.10}$$

The variable factors that fix the design of a true model when the Reynolds number governs are the length ratio and the viscosity ratio.

The *Weber number* is

$$\mathbf{W} = \frac{V^2 L \rho}{\sigma}$$ (10.11)

where ρ = density of fluid, lb $\cdot$ s^2/ft^4 (kg $\cdot$ s^2/m^4) (specific weight divided by g); and σ = surface tension of fluid, lb/ft^2(kPa).

The Weber numbers of model and prototype are equated in certain types of wave studies.

For the flow of water in open channels and rivers where the friction slope is relatively flat, model designs are often based on the Manning equation. The relations between the model and prototype are determined as follows:

$$\frac{V_m}{V_p} = \frac{(1.486/n_m)R_m^{2/3}S_m^{1/2}}{(1.486/n_p)R_p^{2/3}S_p^{1/2}}$$ (10.12)

where n = Manning roughness coefficient ($T/L^{1/3}$, T representing time)
$\quad R$ = hydraulic radius (L)
$\quad S$ = loss of head due to friction per unit length of conduit (dimensionless)
$\quad\ \ $ = slope of energy gradient

For true models, $S_r = 1$ and $R_r = L_r$. Hence,

$$V_r = \frac{L_r^{2/3}}{n_r}$$ (10.13)

In models of rivers and channels, it is necessary for the flow to be turbulent. The U.S. Waterways Experiment Station has determined that flow is turbulent if

$$\frac{VR}{\nu} \geq 4000$$ (10.14)

where V = mean velocity, ft/s (m/s)
$\quad R$ = hydraulic radius, ft (m)
$\quad \nu$ = kinematic viscosity, ft^2/s (m^2/s)

If the model is to be a true model, it may have to be uneconomically large for the flow to be turbulent.

FLUID FLOW IN PIPES

Laminar Flow

In laminar flow, fluid particles move in parallel layers in one direction. The parabolic velocity distribution in laminar flow, shown in Fig. 10.3, creates

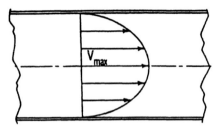

FIGURE 10.3 Velocity distribution for lamellar flow in a circular pipe is parabolic. Maximum velocity is twice the average velocity.

a shearing stress $\tau = \mu \, dV/dy$, where dV/dy is the rate of change of velocity with depth and μ is the coefficient of viscosity. As this shearing stress increases, the viscous forces become unable to damp out disturbances, and turbulent flow results. The region of change is dependent on the fluid velocity, density, and viscosity and the size of the conduit.

A dimensionless parameter called the *Reynolds number* has been found to be a reliable criterion for the determination of laminar or turbulent flow. It is the ratio of inertial forces to viscous forces and is given by

$$\mathbf{R} = \frac{VD\rho}{\mu} = \frac{VD}{\nu} \tag{10.15}$$

where V = fluid velocity, ft/s (m/s)
D = pipe diameter, ft (m)
ρ = density of fluid, lb·s²/ft⁴ (kg·s²/m⁴) (specific weight divided by g, 32.2 ft/s²)
μ = viscosity of fluid lb·s/ft² (kg·s/m²)
ν = μ/ρ = kinematic viscosity, ft²/s (m²/s)

For a Reynolds number less than 2000, flow is laminar in circular pipes. When the Reynolds number is greater than 2000, laminar flow is unstable; a disturbance is probably magnified, causing the flow to become turbulent.

In laminar flow, the following equation for head loss due to friction can be developed by considering the forces acting on a cylinder of fluid in a pipe:

$$h_f = \frac{32\mu LV}{D^2\rho g} = \frac{32\mu LV}{D^2 w} \tag{10.16}$$

where h_f = head loss due to friction, ft (m)
L = length of pipe section considered, ft (m)
g = acceleration due to gravity = 32.2 ft/s² (9.81 m/s²)
w = specific weight of fluid, lb/ft³ (kg/m³)

Substitution of the Reynolds number yields

$$h_f = \frac{64}{\mathbf{R}} \frac{L}{D} \frac{V^2}{2g} \qquad (10.17)$$

For laminar flow, Eq. (10.17) is identical to the Darcy-Weisbach formula because, in laminar flow, the friction $f = 64/\mathbf{R}$.

Turbulent Flow

In turbulent flow, the inertial forces are so great that viscous forces cannot dampen out disturbances caused primarily by the surface roughness. These disturbances create eddies, which have both a rotational and a translational velocity. The translation of these eddies is a mixing action that effects an interchange of momentum across the cross section of the conduit. As a result, the velocity distribution is more uniform, as shown in Fig. 10.4. Experimentation in turbulent flow has shown that

The head loss varies directly as the length of the pipe.

The head loss varies almost as the square of the velocity.

The head loss varies almost inversely as the diameter.

The head loss depends on the surface roughness of the pipe wall.

The head loss depends on the fluid density and viscosity.

The head loss is independent of the pressure.

Darcy-Weisbach Formula

One of the most widely used equations for pipe flow, the Darcy-Weisbach formula satisfies the condition described in the preceding section and is valid for laminar or turbulent flow in all fluids:

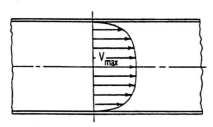

FIGURE 10.4 Velocity distribution for turbulent flow in a circular pipe is more nearly uniform than that for lamellar flow.

$$h_f = f \frac{L}{D} \frac{V^2}{2g} \tag{10.18}$$

where h_f = head loss due to friction, ft (m)
f = friction factor (see an engineering handbook)
L = length of pipe, ft (m)
D = diameter of pipe, ft (m)
V = velocity of fluid, ft/s (m/s)
g = acceleration due to gravity = 32.2 ft/s^2 (9.81 m/s^2)

It employs the Moody diagram for evaluating the friction factor f. (See L. F. Moody, "Friction Factors for Pipe Flow," *Transactions of the American Society of Mechanical Engineers,* November 1944.)

Because Eq. 10.18 is dimensionally homogeneous, it can be used with any consistent set of units without changing the value of the friction factor.

Roughness values ϵ, ft (m), for use with the Moody diagram to determine the Darcy-Weisbach friction factor f are listed in engineering handbooks.

Chezy Formula

This equation holds for head loss in conduits and gives reasonably good results for high Reynolds numbers:

$$V = C \sqrt{RS} \tag{10.19}$$

where V = velocity, ft/s (m/s)
C = coefficient, depends on surface roughness of conduit
S = slope of energy grade line or head loss due to friction, ft/ft (m/m) of conduit
R = hydraulic radius, ft (m)

Hydraulic radius of a conduit is the cross-sectional area of the fluid in it, divided by the perimeter of the wetted section.

Manning's Formula

Through experimentation, Manning concluded that the C in the Chezy equation should vary as $R^{1/6}$:

$$C = \frac{1.486R^{1/6}}{n} \tag{10.20}$$

where n = coefficient, dependent on surface roughness. (Although based on surface roughness, n in practice is sometimes treated as a lumped parameter for all head losses.) Substitution gives

$$V = \frac{1.486}{n} R^{2/3}S^{1/2} \tag{10.21}$$

Upon substitution of $D/4$, where D is the pipe diameter, for the hydraulic radius of the pipe, the following equations are obtained for pipes flowing full:

$$V = \frac{0.590}{n} D^{2/3}S^{1/2} \tag{10.22}$$

$$Q = \frac{0.463}{n} D^{8/3}S^{1/2} \tag{10.23}$$

$$h_f = 4.66n^2 \frac{LQ^2}{D^{16/3}} \tag{10.24}$$

$$D = \left(\frac{2.159Qn}{S^{1/2}}\right)^{3/8} \tag{10.25}$$

where Q = flow, ft³/s (m³/s).

Hazen-Williams Formula

This is one of the most widely used formulas for pipeflow computations of water utilities, although it was developed for both open channels and pipe flow:

$$V = 1.318C_1R^{0.63}S^{0.54} \tag{10.26}$$

For pipes flowing full,

$$V = 0.55C_1D^{0.63}S^{0.54}$$

$$Q = 0.432C_1D^{2.63}S^{0.54} \tag{10.27}$$

$$h_f = \frac{4.727}{D^{4.87}} L \left(\frac{Q}{C_1}\right)^{1.85} \tag{10.28}$$

$$D = \frac{1.376}{S^{0.205}} \left(\frac{Q}{C_1}\right)^{0.38} \tag{10.29}$$

where V = velocity, ft/s (m/s)
C_1 = coefficient, dependent on surface roughness (given in engineering handbooks)
R = hydraulic radius, ft (m)
S = head loss due to friction, ft/ft (m/m) of pipe

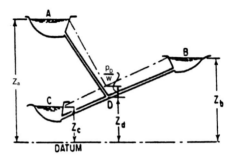

FIGURE 10.5 Flow between reservoirs.

D = diameter of pipe, ft (m)
L = length of pipe, ft (m)
Q = discharge, ft³/s (m³/s)
h_f = friction loss, ft (m)

Figure 10.5 shows a typical three-reservoir problem. The elevations of the hydraulic grader lines for the three pipes are equal at point D. The Hazen-Williams equation for friction loss can be written for each pipe meeting at D. With the continuity equation for quantity of flow, there are as many equations as there are unknowns:

$$Z_a = Z_d + \frac{p_D}{w} + \frac{4.727 L_A}{D_A^{4.87}} \left(\frac{Q_A}{C_A}\right)^{1.85} \qquad (10.30)$$

$$Z_b = Z_d + \frac{p_D}{w} + \frac{4.727 L_B}{D_B^{4.87}} \left(\frac{Q_B}{C_B}\right)^{1.85} \qquad (10.31)$$

$$Z_c = Z_d + \frac{p_D}{w} + \frac{4.727 L_C}{D_C^{4.87}} \left(\frac{Q_C}{C_C}\right)^{1.85} \qquad (10.32)$$

$$Q_A + Q_B = Q_C \qquad (10.33)$$

where p_D = pressure at D and w = unit weight of liquid.

PRESSURE (HEAD) CHANGES CAUSED BY PIPE SIZE CHANGE

Energy losses occur in pipe contractions, bends enlargements, and valves and other pipe fittings. These losses can usually be neglected if the length of the pipeline is greater than 1500 times the pipe diameter. However, in short pipelines, because these losses may exceed the friction losses, minor loses must be considered.

Sudden Enlargements

The following equation for the head loss, ft (m), across a sudden enlargement of pipe diameter has been determined analytically and agrees well with experimental results:

$$h_L = \frac{(V_1 - V_2)^2}{2g} \tag{10.34}$$

where V_1 = velocity before enlargement, ft/s (m/s)
V_2 = velocity after enlargement, ft/s (m/s)
g = 32.2 ft/s^2 (9.81 m/s^2)

Another equation for the head loss caused by sudden enlargements was determined experimentally by Archer. This equation gives slightly better agreement with experimental results than Eq. (10.34):

$$h_L = \frac{1.1(V_1 - V_2)^{1.92}}{2g} \tag{10.35}$$

Gradual Enlargements

The equation for the head loss due to a gradual conical enlargement of a pipe takes the form

$$h_L = \frac{K(V_1 - V_2)^2}{2g} \tag{10.36}$$

where K = loss coefficient, as given in engineering handbooks.

Sudden Contraction

The following equation for the head loss across a sudden contraction of a pipe was determined by the same type of analytic studies:

$$h_L = \left(\frac{1}{C_c} - 1\right)^2 \frac{V^2}{2g} \tag{10.37}$$

where C_c = coefficient of contraction and V = velocity in smaller-diameter pipe, ft/s (m/s). This equation gives best results when the head loss is greater than 1 ft (0.3 m).

Another formula for determining the loss of head caused by a sudden contraction, determined experimentally by Brightmore, is

$$h_L = \frac{0.7(V_1 - V_2)^2}{2g} \tag{10.38}$$

This equation gives best results if the head loss is less than 1 ft (0.3 m).

A special case of sudden contraction is the entrance loss for pipes. Some typical values of the loss coefficient K in $h_L = KV^2/2g$, where V is the velocity in the pipe, are presented in engineering handbooks.

Bends and Standard Fitting Losses

The head loss that occurs in pipe fittings, such as valves and elbows and at bends is given by

$$h_L = \frac{KV^2}{2g} \tag{10.39}$$

To obtain losses in bends other than 90°, the following formula may be used to adjust the K values:

$$K' = K\sqrt{\frac{\Delta}{90}} \tag{10.40}$$

where Δ = deflection angle in degrees. The K values are given in engineering handbooks.

FLOW THROUGH ORIFICES

An orifice is an opening with a closed perimeter through which water flows. Orifices may have any shape, although they are usually round, square, or rectangular.

Orifice Discharge into Free Air

Discharge through a sharp-edged orifice may be calculated from

$$Q = Ca\sqrt{2gh} \tag{10.41}$$

where Q = discharge ft^3/s (m^3/s)
   C = coefficient of discharge
   a = area of orifice, ft^2 (m^2)

g = acceleration due to gravity, ft/s² (m/s²)
h = head on horizontal centerline of orifice, ft (m)

Coefficients of discharge C are given in engineering handbooks for low velocity of approach. If this velocity is significant, its effect should be taken into account. For low heads, measuring the head from the centerline of the orifice is not theoretically correct; however, this error is corrected by the C values.

The *coefficient of discharge* C is the product of the coefficient of velocity C_v and the coefficient of contraction C_c. The *coefficient of velocity* is the ratio obtained by dividing the actual velocity at the *vena contracta* (contraction of the jet discharged) by the theoretical velocity. The theoretical velocity may be calculated by writing Bernoulli's equation for points 1 and 2 in Fig. 10.6 as

$$\frac{V_1^2}{2g} + \frac{p_1}{w} + Z_1 = \frac{V_2^2}{2g} + \frac{p_2}{w} + z_2 \qquad (10.42)$$

With the reference plane through point 2, $Z_1 = h$, $V_1 = 0$, $p_1/w = p_2/w = 0$, and $Z_2 = 0$, so Eq. (10.42) becomes

$$V_2 = \sqrt{2gh} \qquad (10.43)$$

The *coefficient of contraction* C_c is the ratio of the smallest area of the jet, the vena contracta, to the area of the orifice. Contraction of a fluid jet occurs if the orifice is square-edged and so located that some of the fluid approaches the orifice at an angle to the direction of flow through the orifice.

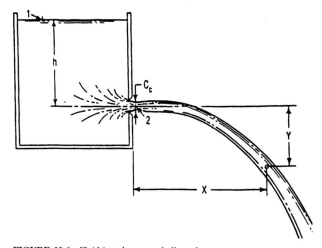

FIGURE 10.6 Fluid jet takes a parabolic path.

Submerged Orifices

Flow through a submerged orifice may be computed by applying Bernoulli's equation to points 1 and 2 in Fig. 10.7.

$$V_2 = \sqrt{2g \left(h_1 - h_2 + \frac{V_1^2}{2g} - h_L \right)} \qquad (10.44)$$

where h_L = losses in head, ft (m), between points 1 and 2.

By assuming $V_1 \approx 0$, setting $h_1 - h_2 = \Delta h$, and using a coefficient of discharge C to account for losses, the following formula is obtained:

$$Q = Ca \sqrt{2g \, \Delta h} \qquad (10.45)$$

Values of C for submerged orifices do not differ greatly from those for nonsubmerged orifices.

Discharge under Falling Head

The flow from a reservoir or vessel when the inflow is less than the outflow represents a condition of falling head. The time required for a certain quantity of water to flow from a reservoir can be calculated by equating the volume of water that flows through the orifice or pipe in time dt to the volume decrease in the reservoir. If the area of the reservoir is constant,

$$t = \frac{2A}{Ca \sqrt{2g}} (\sqrt{h_1} - \sqrt{h_2}) \qquad (10.46)$$

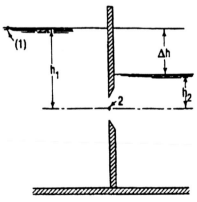

FIGURE 10.7 Discharge through a submerged orifice.

where h_1 = head at the start, ft (m)
h_2 = head at the end, ft (m)
t = time interval for head to fall from h_1 to h_2, s

FLUID JETS

Where the effect of air resistance is small, a fluid discharged through an orifice into the air follows the path of a projectile. The initial velocity of the jet is

$$V_0 = C_v\sqrt{2gh} \tag{10.47}$$

where h = head on centerline of orifice, ft (m), and C_v coefficient of velocity.

The direction of the initial velocity depends on the orientation of the surface in which the orifice is located. For simplicity, the following equations were determined assuming the orifice is located in a vertical surface (see Fig. 10.6). The velocity of the jet in the X direction (horizontal) remains constant:

$$V_x = V_0 = C_v\sqrt{2gh} \tag{10.48}$$

The velocity in the Y direction is initially zero and thereafter a function of time and the acceleration of gravity:

$$V_y = gt \tag{10.49}$$

The X coordinate at time t is

$$X = V_x t = t C_v\sqrt{2gh} \tag{10.50}$$

The Y coordinate is

$$Y = V_{avg}t = \frac{gt^2}{2} \tag{10.51}$$

where V_{avg} = average velocity over period of time t. The equation for the path of the jet is

$$X^2 = C_v^2 4hY \tag{10.52}$$

ORIFICE DISCHARGE INTO DIVERGING CONICAL TUBES

This type of tube can greatly increase the flow through an orifice by reducing the pressure at the orifice below atmospheric. The formula that follows

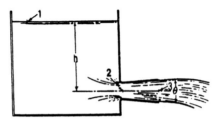

FIGURE 10.8 Diverging conical tube increases flow from a reservoir through an orifice by reducing the pressure below atmospheric.

for the pressure at the entrance to the tube is obtained by writing the Bernoulli equation for points 1 and 3 and points 1 and 2 in Fig. 10.8:

$$p_2 = wh \left[1 - \left(\frac{a_3}{a_2} \right)^2 \right] \qquad (10.53)$$

where p_2 = gauge pressure at tube entrance, lb/ft² (Pa)
w = unit weight of water, lb/ft³ (kg/m³)
h = head on centerline of orifice, ft (m)
a_2 = area of smallest part of jet (vena contracta, if one exists), ft² (m)
a_3 = area of discharge end of tube, ft² (m²)

Discharge is also calculated by writing the Bernoulli equation for points 1 and 3 in Fig. 10.8.

For this analysis to be valid, the tube must flow full, and the pressure in the throat of the tube must not fall to the vapor pressure of water. Experiments by Venturi show the most efficient angle θ to be around 5°.

WATER HAMMER

Water hammer is a change in pressure, either above or below the normal pressure, caused by a variation of the flow rate in a pipe.

The equation for the velocity of a wave in a pipe is

$$U = \sqrt{\frac{E}{\rho}} \sqrt{\frac{1}{1 + ED/E_p t}} \qquad (10.54)$$

where U = velocity of pressure wave along pipe, ft/s (m/s)
E = modulus of elasticity of water = 43.2×10^6 lb/ft² (2.07×10^6 kPa)

ρ = density of water = 1.94 lb·s/ft^4 (specific weight divided by acceleration due to gravity)

D = diameter of pipe, ft (m)

E_p = modulus of elasticity of pipe material, lb/ft^2 (kg/m^2)

t = thickness of pipe wall, ft (m)

PIPE STRESSES PERPENDICULAR TO THE LONGITUDINAL AXIS

The stresses acting perpendicular to the longitudinal axis of a pipe are caused by either internal or external pressures on the pipe walls.

Internal pressure creates a stress commonly called *hoop tension*. It may be calculated by taking a free-body diagram of a 1-in- (25.4-mm)-long strip of pipe cut by a vertical plane through the longitudinal axis (Fig. 10.9). The forces in the vertical direction cancel. The sum of the forces in the horizontal direction is

$$pD = 2F \qquad (10.55)$$

where p = internal pressure, lb/in^2 (MPa)

D = outside diameter of pipe, in (mm)

F = force acting on each cut of edge of pipe, lb (N)

Hence, the stress, lb/in^2 (MPa), on the pipe material is

$$f = \frac{F}{A} = \frac{pD}{2t} \qquad (10.56)$$

where A = area of cut edge of pipe, ft^2 (m^2), and t = thickness of pipe wall, in (mm).

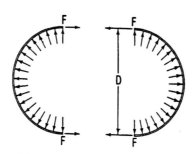

FIGURE 10.9 Internal pipe pressure produces hoop tension.

TEMPERATURE EXPANSION OF PIPE

If a pipe is subject to a wide range of temperatures, the stress, lb/in^2 (MPa), due to a temperature change is

$$f = cE \, \Delta T \qquad (10.57)$$

where E = modulus of elasticity of pipe material, lb/in^2 (MPa)
ΔT = temperature change from installation temperature
c = coefficient of thermal expansion of pipe material

The movement that should be allowed for, if expansion joints are to be used, is

$$\Delta L = Lc \, \Delta T \qquad (10.58)$$

where ΔL = movement in length L of pipe and L = length between expansion joints.

FORCES DUE TO PIPE BENDS

It is common practice to use thrust blocks in pipe bends to take the forces on the pipe caused by the momentum change and the unbalanced internal pressure of the water.

The force diagram in Fig. 10.10 is a convenient method for finding the resultant force on a bend. The forces can be resolved into X and Y components to find the magnitude and direction of the resultant force on the pipe. In Fig. 10.10,

V_1 = velocity before change in size of pipe, ft/s (m/s)

V_2 = velocity after change in size of pipe, ft/s (m/s)

p_1 = pressure before bend or size change in pipe, lb/ft^2 (kPa)

p_2 = pressure after bend or size change in pipe, lb/ft^2 (kPa)

A_1 = area before size change in pipe, ft^2 (m^2)

A_2 = area after size change in pipe, ft^2 (m^2)

F_{2m} = force due to momentum of water in section 2 = $V_2 Qw/g$

F_{1m} = force due to momentum of water in section 1 = $V_1 Qw/g$

P_2 = pressure of water in section 2 times area of section 2 = $p_2 A_2$

P_1 = pressure of water in section 1 times area of section 1 = $p_1 A_1$

w = unit weight of liquid, lb/ft^3 (kg/m^3)

Q = discharge, ft^3/s (m^3/s)

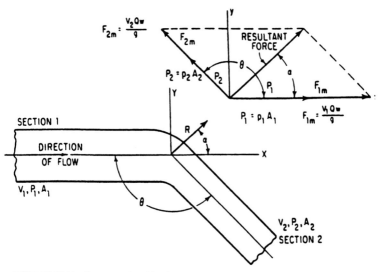

FIGURE 10.10   Forces produced by flow at a pipe bend and change in diameter.

If the pressure loss in the bend is neglected and there is no change in magnitude of velocity around the bend, a quick solution is

$$R = 2A \left(w \frac{V^2}{g} + p \right) \cos \frac{\theta}{2} \qquad (10.59)$$

$$\alpha = \frac{\theta}{2} \qquad (10.60)$$

where R = resultant force on bend, lb (N)
α = angle that R makes with F_{1m}
p = pressure, lb/ft^2 (kPa)
w = unit of weight of water = 62.4 lb/ft^3 (998.4 kg/m^3)
V = velocity of flow, ft/s (m/s)
g = acceleration due to gravity = 32.2 ft/s^2 (9.81 m/s^2)
A = area of pipe, ft^2 (m^2)
θ = angle between pipes (0° ≤ θ ≤ 180°)

ECONOMICAL SIZING OF DISTRIBUTION PIPING

An equation for the most economical pipe diameter for a distribution system for water is

$$D = 0.215 \left(\frac{fbQ_a^3 S}{aiH_a}\right)^{1/7} \qquad (10.61)$$

where D = pipe diameter, ft (m)
$\quad f$ = Darcy-Weisbach friction factor
$\quad b$ = value of power, $/hp per year ($/kW per year)
$\quad Q_a$ = average discharge, ft³/s (m/s)
$\quad S$ = allowable unit stress in pipe, lb/in² (MPa)
$\quad a$ = in-place cost of pipe, $/lb ($/kg)
$\quad i$ = yearly fixed charges for pipeline (expressed as a fraction of total capital cost)
$\quad H_a$ = average head on pipe, ft (m)

DETERMINING DIAMETER NEEDED FOR STEAM AND WATER PIPING

Calculation of fluid losses is necessary to determine precise pipe size. Hence this is the starting point in any piping system design—whether for liquid or gas. It is usually best economically to keep velocities as high as possible consistent with reasonable losses. Optimum pipe size will give minimum annual charges—pumping cost plus capitalized installation cost.

Start out by picking a velocity from table below, depending on the application. But remember that it will only be approximate, to be rechecked later in your calculation to arrive at best economic size.

Service	Velocity range ft/s	m/s
Service water mains	2 to 5	0.61 to 1.5
General-service water piping	4 to 10	1.2 to 3.1
Boiler feedwater piping	6 to 18	1.8 to 5.5
Low-pressure steam-heating piping	15 to 70	4.6 to 21.3
Low pressure steam mains	70 to 165	21.3 to 50.3
High-pressure steam mains	165 to 400	50.3 to 121.9
Steam-engine and pump piping	100 to 150	30.5 to 45.7
Steam-turbine piping	150 to 330	45.7 to 100.6

Allowable velocity increases somewhat with pipe diameter, so recheck velocity V from the equations below, using nominal pipe diameter d:

Pump discharge lines, ft/s	$V = (d/2) + 4$	(10.62)
Pump suction lines, ft/s	$V = 1/3 \ (d/2) + 4$	(10.63)
Steam lines, 1000 ft/min	$V = d$	(10.64)

Given the approximate velocity, solve for pipe size:

$$\text{ID of pipe} = \sqrt{\frac{0.409 \times \text{gal/min}}{\text{velocity, ft/s}}} \qquad (10.65)$$

VENTURI METER FLOW COMPUTATION

Flow through a venturi meter (Fig. 10.11) is given by

$$Q = cKd_2^2 \sqrt{h_1 - h_2} \qquad (10.66)$$

$$K = \frac{4}{\pi} \sqrt{\frac{2g}{1 - (d_2/d_1)^2}} \qquad (10.67)$$

where Q = flow rate, ft³/s (m³/s)
  c = empirical discharge coefficient dependent on throat velocity and diameter
  d_1 = diameter of main section, ft (m)
  d_2 = diameter of throat, ft (m)
  h_1 = pressure in main section, ft (m) of water
  h_2 = pressure in throat section, ft (m) of water

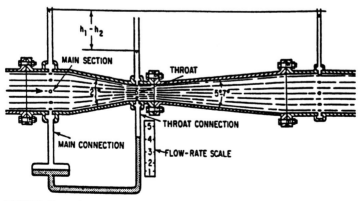

FIGURE 10.11   Standard venturi meter.

NONUNIFORM FLOW IN OPEN CHANNELS

Symbols used in this section are V = velocity of flow in the open channel, ft/s (m/s); D_c = critical depth, ft (m); g = acceleration due to gravity, ft/s^2 (m/s^2); Q = flow rate, ft^3/s (m^3/s); q = flow rate per unit width, ft^3/ft (m^3/m); H_m = minimum specific energy, ft · lb/lb (kg · m/kg). Channel dimensions are in feet or meters and the symbols for them are given in the text and illustrations.

Nonuniform flow occurs in open channels with gradual or sudden changes in the cross-sectional area of the fluid stream. The terms *gradually varied flow* and *rapidly varied flow* are used to describe these two types of nonuniform flow. Equations are given next for flow in (1) rectangular cross-section channels, (2) triangular channels, (3) parabolic channels, (4) trapezoidal channels, and (5) circular channels. These five types of channels cover the majority of actual examples met in the field. Figure 10.12 shows the general energy relations in open-channel flow.

Rectangular Channels

In a rectangular channel, the critical depth D_c equals the mean depth D_m; the bottom width of the channel b equals the top width T; and when the discharge of fluid is taken as the flow per foot (meter) of width q of the channel, both b and T equal unity. Then the average velocity V_c is

$$V_c = \sqrt{gD_c} \tag{10.68}$$

and

$$D_c = \frac{V_c^2}{g} \tag{10.69}$$

Also

$$Q = \sqrt{g}\, bD_c^{3/2} \tag{10.70}$$

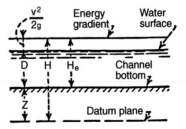

FIGURE 10.12 Energy of open-channel fluid flow.

where g = acceleration due to gravity in USCS or SI units.

$$q = \sqrt{g}D_c^{3/2} \qquad (10.71)$$

and

$$D_c = \sqrt[3]{\frac{q^2}{g}} \qquad (10.72)$$

The minimum specific energy is

$$H_m = \frac{3}{2}D_c \qquad (10.73)$$

and the critical depth is

$$D_c = \tfrac{2}{3}H_m \qquad (10.74)$$

Then the discharge per foot (meter) of width is given by

$$q = \sqrt{g}(\tfrac{2}{3})^{3/2}H_m^{3/2} \qquad (10.75)$$

With $g = 32.16$, Eq. (10.75) becomes

$$q = 3.087H_m^{3/2} \qquad (10.76)$$

Triangular Channels

In a triangular channel (Fig. 10.13), the maximum depth D_c and the mean depth D_m equal $\tfrac{1}{2}D_c$. Then

$$V_c = \sqrt{\frac{gD_c}{2}} \qquad (10.77)$$

and

$$D_c = \frac{2V_c^2}{g} \qquad (10.78)$$

As shown in Fig. 10.13, z is the slope of the channel sides, expressed as a ratio of horizontal to vertical; for symmetric sections, $z = e/D_c$. The area $a = zD_c^2$. Then

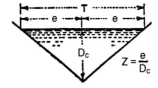

FIGURE 10.13 Triangular open channel.

$$Q = \sqrt{\frac{g}{2}}\, z D_c^{5/2} \qquad (10.79)$$

with $g = 32.16$,

$$Q = 4.01 z D_c^{5/2} \qquad (10.80)$$

and

$$D_c = \sqrt{\frac{2Q^2}{gz^2}} \qquad (10.81)$$

or

$$Q = \sqrt[5]{\frac{g}{2}}\left(\frac{4}{5}\right)^{5/2} z H_m^{5/2} \qquad (10.82)$$

With $g = 32.16$,

$$Q = 2.295 z H_m^{5/2} \qquad (10.83)$$

Parabolic Channel

These channels can be conveniently defined in terms of the top width T and the depth D_c. Then the area $a = \frac{2}{3} D_c T$ and the mean depth is D_m.
Then (Fig. 10.14)

$$V_c = \sqrt{\frac{2}{3} g D_c} \qquad (10.84)$$

and

$$D_c = \frac{3}{2}\frac{V_c^2}{g} \qquad (10.85)$$

Further,

$$Q = \sqrt{\frac{8g}{27}}\, T D_c^{3/2} \qquad (10.86)$$

With $g = 32.16$,

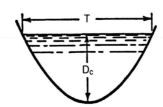

FIGURE 10.14 Parabolic open channel.

$$Q = 3.087TD_c^{3/2} \qquad (10.87)$$

and
$$D_c = \frac{3}{2} \sqrt[3]{\frac{Q^2}{gT^2}} \qquad (10.88)$$

Also
$$Q = \sqrt{\frac{8g}{27}} \left(\frac{3}{4}\right)^{3/2} TH_m^{3/2} \qquad (10.89)$$

With $g = 32.16$,

$$Q = 2.005TH_m^{3/2} \qquad (10.90)$$

Trapezoidal Channels

Figure 10.15 shows a trapezoidal channel having a depth of D_c and a bottom width b. The slope of the sides, horizontal divided by vertical, is z. By expressing the mean depth D_m in terms of channel dimensions, the relations for critical depth D_c and average velocity V_c are

$$V_c = \sqrt{\frac{b + zD_c}{b + 2zD_c} gD_c} \qquad (10.91)$$

and
$$D_c = \frac{V_c^2}{c} - \frac{b}{2z} + \sqrt{\frac{V_c^4}{g^2} + \frac{b^2}{4z^2}} \qquad (10.92)$$

The discharge through the channel is

$$Q = \sqrt{g \frac{(b + zD_c)^3}{b + 2zD_c}} D_c^{3/2} \qquad (10.93)$$

Then the minimum specific energy and critical depth are

$$H_m = \frac{3b + 5zD_c}{2b + 4zD_c} D_e \qquad (10.94)$$

$$D_c = \frac{4zH_m - 3b + \sqrt{16z^2H_m^2 + 16zH_mb + 9b^2}}{10z} \qquad (10.95)$$

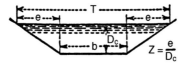

FIGURE 10.15   Trapezoidal open channel.

Circular Channels

Figure 10.16 shows a typical circular channel in which the area a, top width T, and depth D_c are

$$a = \frac{d^2}{4}\left(\theta_r - \frac{1}{2}\sin 2\theta\right)$$ (10.96)

$$T = d \sin \theta$$ (10.97)

$$D_c = \frac{d}{2}(1 - \cos \theta)$$ (10.98)

Flow quantity is then given by

$$Q = \frac{2^{3/2}g^{1/2}(\theta_r - \frac{1}{2}\sin 2\theta)^{3/2}}{8(\sin \theta)^{1/2}(1 - \cos \theta)^{5/2}} D_c^{5/2}$$ (10.99)

PUMPS

Definitions

A *pump* is a machine or device for raising a liquid, which is a relatively incompressible fluid, to a higher level or to a higher pressure. A *compressor* is a machine or device for raising a gas, which is a compressible fluid, to a higher pressure. However, devices for exhausting air from closed vessels are called *air pumps,* although in reality they are air compressors working below atmospheric pressure.

A *blower,* as distinguished from a compressor, compresses a gas to a comparatively low pressure only. A *fan* is intended primarily to move large volumes of gas; the pressure developed by the fan is quite small and is secondary in importance.

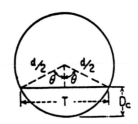

FIGURE 10.16   Circular channel.

Measurement of Head

The head that a pump has to develop or work against is the static lift plus all the friction losses in the piping. This value may be computed, but in actual operation it would be determined in a test by measuring the pressures in the piping adjacent to the pump on both suction and discharge sides. Let h = total head in feet (meters), p = pressure expressed in feet (meters) of the liquid, z = elevation of the center of the discharge gauge above the point at which the suction pressure is measured, V = velocity in feet per second (meters per second) at the section where the gauge is attached, g = acceleration of gravity in feet per square second (meters per square second), the subscript d denotes discharge, and the subscript s denotes suction values. Then

$$h = p_d - p_s + z + \frac{V_d^2}{2g} - \frac{V_s^2}{2g} \qquad (10.100)$$

If the pressure on the intake side is below atmospheric and if gauge pressures are used in the above equation, then p_s will be negative.

Power

If q = rate of discharge, ft³/s (m³/s), G = gal/min (L/min), w = density of the liquid, lb/ft³ (kg/m³), the horsepower (W) delivered in the liquid called *water horsepower,* is

$$\text{Water power} = \frac{wqh}{550} \text{ hp} \qquad (1 \text{ hp} = 0.75 \text{ kW})$$

In the case of water of the customary density of 62.4 lb/ft³ (8.0 kg/m³), this may be reduced to

$$\text{Water power} = \frac{qh}{8.81} = \frac{Gh}{3960} \quad \text{hp} \qquad (1 \text{ hp} = 0.75 \text{ kW}) \quad (10.101)$$

For any other liquid of specific gravity s, the two expressions in Eq. (10.101), and in Eq. (10.102), should be multiplied by s. If e is the overall efficiency of the pump, then the power input to the pump, often called *brake horsepower,* is

$$\text{Brake power} = \frac{qh}{e \times 8.81} = \frac{Gh}{e \times 3960} \quad \text{hp} \qquad (1 \text{ hp} = 0.75 \text{ kW})$$

$$(10.102)$$

Efficiencies Defined

Efficiency, sometimes called *total* or *overall efficiency,* is the ratio of the power delivered in the liquid to the power input to the pump. That is,

$$e = \frac{\text{water hp}}{\text{brake hp}} = \frac{\text{water kW}}{\text{brake kW}} \qquad (10.103)$$

Hydraulic efficiency e_h is the ratio of the power actually delivered *in* the water to the power expended *on* the water or other liquid. These two quantities differ by the amount of the hydraulic friction losses.

Mechanical efficiency e_m is the ratio of the power expended *on* the water to the power supplied to run the pump. These two differ by the amount of the mechanical friction losses, such as friction of bearings, stuffing boxes, etc.

Volumetric efficiency e_v is the ratio of the amount of water actually delivered to that which would be delivered if there were no leakage losses, imperfect valve action, etc. *Slip,* in the case of a positive-displacement pump, means the difference between the actual displacement and the volume of the fluid actually delivered, expressed as a percentage of the displacement. The relation between slip and volumetric efficiency is: slip $= 100(1 - e_v)$.

The *total efficiency* is the product of the hydraulic, mechanical, and volumetric efficiencies, that is,

$$e = e_h \times e_m \times e_v \qquad (10.104)$$

Duty is another means of expressing the efficiency of steam-driven pumping engines. It is usually expressed as the foot-pounds of work done per 1000 lb (J/454 kg) of steam supplied, but is more precisely defined as the foot-pounds of work done per million Btu (J/1.1 MJ) supplied.

Suction Lift

The theoretical suction lift may be computed as follows:

$$L = b - p_v - h_f - \frac{V_s^2}{2g} \qquad (10.105)$$

where L = lift; b = barometer pressure, ft (m), of the liquid; p_v = vapor pressure of the liquid, ft (m); h_f = friction losses in foot valve, suction piping, etc; and V_s = velocity at intake of pump. For water it is desirable to maintain a pressure at least about 10 ft (3 m) more than the vapor pressure. Hence, the maximum allowable lift is about 10 ft (3 m) less than given by Eq. (10.105). In practice the lift is usually about 20 ft (6 m) for cold water, decreasing as the water temperature increases; above 160°F (71°C), the water should be supplied under a positive pressure.

ENERGY IN PUMPING SYSTEMS

An incompressible fluid has energy in the form of velocity, pressure, and elevation. Bernoulli's theorem for an incompressible fluid states that in steady flow without losses, the energy at any point is the sum of the velocity head, pressure head, and elevation head, and that this sum is constant along a stream line in the conduit. Therefore the energy H, ft·lb/lb, or ft(m), gauge or absolute, at any point in the system relative to a selected datum plane is

$$H = \frac{V^2}{2g} + \frac{144p}{w} + Z \qquad (10.106)$$

where V = velocity, ft/s (m/s)
g = acceleration of gravity, approximately 32.17 ft/s² (9.81 m/s²)
p = pressure (+ or −), lb/in² gauge or abs (kPa)
w = specific weight of liquid, lb/ft³ (kg/m³)
Z = elevation above (+) or (−) datum, ft (m)

PUMP TOTAL HEAD

The total head of a pump is the difference in energy between the pump discharge (point 2) and the pump suction (point 1), as shown in Fig. 10.17. Applying Bernoulli's theorem at each point, Eq. (10.7), the pump total head TH in feet (meters) becomes

$$\text{TH} = H_d - H_s = \left(\frac{V_d^2}{2g} + \frac{144p_d}{w_d} + Z_d \right) - \left(\frac{V_s^2}{2g} + \frac{144p_s}{w_s} + Z_s \right)$$

$$(10.107)$$

This equation for pump differential pressure P_Δ, psi (kPa), is

$$P_\Delta = P_d - P_s$$

$$= \left[p_d + 0.433 \text{ sp gr}_d \left(Z_d + \frac{V_d^2}{2g} \right) \right]$$

$$- \left[p_s + 0.433 \text{ sp gr}_s \left(Z_s + \frac{V_s^2}{2g} \right) \right] \qquad (10.108)$$

where for Eqs. (10.107) and (10.108) subscripts d and s denote discharge and suction, respectively, and

H = total head, (+ or −) ft gauge (or ft abs) (m)
P = total pressure, (+ or −) psi gauge (or psi abs) (kPa)

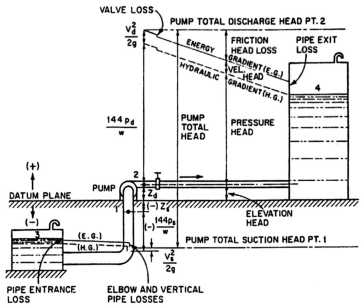

FIGURE 10.17 Energy and hydraulic gradients.

V = velocity, ft/s (m/s)

p = pressure, (+ or −) psi gauge (or psi abs) (kPa)

Z = elevation above (+) or below (−) datum, ft (m)

w = specific weight of liquid, lb/ft³ (kg/m³)

sp gr = specific gravity of liquid

g = acceleration of gravity, approx. 32.17 ft/s² (9.81 m/s²)

Pump total head TH, ft, and pump differential pressure $P\Delta$, psi (kPa), are always absolute quantities since either gauge pressures or absolute pressures but not both are used at the discharge and suction connections of the pump, and a common datum plane is selected.

Pump total head in feet (meters) and pump differential pressure, psi (kPa) are related to each other as follows:

$$\text{TH} = \frac{144 P\Delta}{w} \qquad (10.109)$$

Between any two points in a pumping system where the energy is added only by the pump and the specific weight of the liquid does not change (e.g., as a result of temperature), the following general equation for determining pump total head applies:

$$TH = H_2 - H_1 + \Sigma h_{f(1-2)}$$
$$= \left(\frac{V_2^2}{2g} + \frac{144p_2}{w} + Z_2 \right) - \left(\frac{V_1^2}{2g} + \frac{144p_1}{w} + Z_1 \right) + \Sigma h_{f(1-2)}$$

$$(10.110)$$

where subscripts 1 and 2 denote points in the pumping system anyplace upstream and downstream from the pump, respectively, and

H = total head (+ or −), ft gauge (or ft abs) (m)
V = velocity, ft/s (m/s)
p = pressure, (+ or −) lb/in^2 gauge (or psi abs) (kPa)
Z = elevation above (+) or below (−) datum, ft (m)
w = specific weight of liquid (assumed the same between points), lb /ft^3 (kg/m^3)
g = acceleration of gravity, approx. 32.17 ft/s^2 (9.81 m/s^2)
Σh_f = sum of the losses between points, ft (m)

When the specific gravity of the liquid is known, the pressure head in feet (meters) may be calculated from the following relationship:

$$\frac{144p}{w} = \frac{2.31p}{sp\ gr} \qquad (10.111)$$

The velocity in feet per second (meters per second) in a pipe may be calculated as follows:

$$V = \frac{gpm \times 0.408}{(pipe\ ID)^2} \qquad (10.112)$$

PUMP FLOW, POWER, AND PRESSURE FORMULAS

The flow rate developed by a rotating pump gpm (gallons per minute) varies as the rpm (revolutions per minute) of the shaft, or

$$\frac{gpm_2}{gpm_1} = \frac{rpm_2}{rpm_1} \qquad (10.113)$$

where the subscripts 1 and 2 represent two different flow rates and shaft rotational speeds. Head developed HD by the pump varies as the square of the flow rate and square of the shaft rotation speed, Or

$$\frac{HD_2}{HD_1} = \left(\frac{gpm_2}{gpm_1}\right)^2 = \left(\frac{rpm_2}{rpm_1}\right)^2 \qquad (10.114)$$

Power input to the pump varies as the cube of the flow rate, cube of the shaft rotation speed, and 1.5 power of the head developed. Or

$$\frac{bhp_2}{bhp_1} = \left(\frac{gpm_2}{gpm_1}\right)^3 = \left(\frac{rpm_2}{rpm_1}\right)^3 = \left(\frac{HD_2}{HD_1}\right)^{1.5} \qquad (10.115)$$

The brake horsepower input to the pump bhp, with a given flow rate gpm and a specific gravity of the fluid pumped of sp gr, and a pump efficiency of Pump$_{EFF}$, expressed as a decimal, is

$$bhp = \frac{gpm \times HD \times sp\ gr}{3960 \times pump_{EFF}} \qquad (10.116)$$

With an electrically driven pump the motor horsepower required to drive the pump is, with a motor/drive (M/D_{EFF}) of 85 to 95 percent,

$$mhp = \frac{bhp}{M/D_{EFF}} \qquad (10.117)$$

The velocity head VH, ft (m), in a pumping system is given by

$$VH = \frac{V^2}{2g} \qquad (10.118)$$

where V = velocity of the liquid in the pipe, ft/s (m/s), and g = acceleration due to gravity = 32.2 ft/s^2 (9.81 m/s^2).

Pump head, in feet (meters) of liquid being pumped, is given by

$$HD = \frac{P \times 2.31}{sp\ gr} \qquad (10.119)$$

where sp gr = specific gravity of the liquid being pumped, compared to sp gr of water = 1.0.

PRESSURE LOSSES IN PIPING AND FITTINGS

When fluid flows through a pipe and fittings, there is a pressure loss caused by the friction in the pipe and the resistance in the fittings. In long pipelines the pressure loss caused by bends, valves, and fittings is often considered to be negligible compared to the loss produced by the length of the piping itself. In short piping systems, with little straight piping, the pressure loss caused by bends, valves, and fittings can be significant.

For the pressure loss in straight piping, use the Darcy-Weisbach formula (10.18), the Manning formula (10.20), or the Hazen-Williams formula (10.26). Where bends, valves, and fittings are present in the pipeline, as is often the case, use the equivalent length for each element, as shown in Table 10.2. The equivalent length is added to the total pipeline length to determine the overall length to be used in the pressure loss computation. Any of the above-named formulas can be used in the computation after the equivalent length is added to the total length.

NET POSITIVE SUCTION HEAD FOR RECIPROCATING PUMPS

With a reciprocating pump the piston moves back and forth in the cylinder, creating an acceleration head. Hence, we must deal with the net positive suction head available.

Net Positive Suction Head Available (NPSHA)

NPSHA is the static head plus atmospheric head, minus lift loss, friction loss, vapor pressure, velocity head, and acceleration head in feet available at the suction connection centerline.

Acceleration head can be the highest factor of NPSHA. In some cases it is 10 times the total of all the other losses. Data from both the pump and the suction system are required to determine acceleration head; its value cannot be calculated until these data have been established.

Acceleration Head

The flow in the suction line is always fluctuating, continuously accelerating or decelerating. The acceleration head is not a loss, because the energy is restored during deceleration. *Acceleration head* is defined as

$$H_a = \frac{L \times V \times n \times C}{g \times K} \qquad (10.120)$$

where H_a = acceleration head loss, ft (m)
L = length of suction pipe, ft (m)

TABLE 10.2 Representative Equivalent Length in Pipe Diameters L/D of Various Valves and Fittings

Globe valves, fully open	450
Angle valves, fully open	200
Gate valves, fully open	13
Three-fourths open	35
One-half open	160
One-fourth open	900
Swing check valves, fully open	135
In line, ball check valves, fully open	150
Butterfly valves, 6-in and larger, fully open	20
90° standard elbow	30
45° standard elbow	16
90° long-radius elbow	20
90° street elbow	50
45° street elbow	26
Standard tee:	
Flow through run	20
Flow through branch	60

Source: Compiled from data given in *Flow of Fluids*, Crane Company Technical Paper 410, ASME, 1971.

V = mean velocity in the suction pipe, ft/s (m/s)

$$V = \frac{gpm \times 0.321}{\text{area of the suction pipe, in}^2} \qquad (10.121)$$

n = pump rotation speed, rpm
g = acceleration of gravity = 32.2 ft/s^2
K = factors for the various fluids:

Water	1.4
Petroleum	2.5
Liquid with entrained gas	1.0

C = factor for the type of pump:

Duplex	0.115
Triplex	0.006
Quadruplex	0.08
Quintuplex	0.04
Sextuplex	0.055
Septuplex	0.028

When the suction system consists of pipes of various sizes, calculate the acceleration head for each section separately. Add the acceleration head of all sections to obtain the total.

If the calculated NPSHA, including acceleration head, is greater than the suction system can provide, the system NPSH should be increased. This can be accomplished by

1. Increasing the static head
2. Increasing the atmospheric pressure
3. Adding a booster pump to the system
4. Adding a pulsation dampener

The basic definition of acceleration head is

$$G_s = \begin{matrix} V \times n \times C = \text{acceleration of} \\ \text{liquid in suction line, ft/s}^2 \end{matrix} \qquad (10.122)$$

$$F_s = \frac{W_s \times G_s}{g} = \text{force to produce acceleration, lb} \qquad (10.123)$$

where W_s = weight of liquid in line = LA, × sp gr, and   (10.124)

$$H_t = \frac{F_s \times 2.31}{A_s \times \text{sp gr}} = \text{theoretical head, ft of liquid} \qquad (10.125)$$

where A_s = cross-sectional area of pipe. Substituting, we get

$$H_L = \frac{W_s VnC \times 2.31}{A_s \times \text{sp gr} \times g} = \frac{LVnC}{g} \qquad (10.126)$$

$$H_a = \frac{Ht}{K} \qquad (10.127)$$

where K = ratio of theoretical head to the actual head. Therefore

$$H_a = \frac{LVnC}{gK} \qquad (10.128)$$

where rpm = pump speed, rpm
h_s = suction head, ft (m)
h_f = piping friction loss, ft (m)
A_s = area of suction pipe, in^2 (mm^2)
L = length of connecting rod, C to C, ft (m)
R = crank radius, ft (m)
l = length of pipe where resistance of flow is to be measured, ft (m)
A_p = area of plunger, in^2 (mm^2)

264   SECTION TEN

NET POSITIVE SUCTION HEAD FOR
CENTRIFUGAL PUMPS

The *net positive suction head* (NPSH) h_{sv} is a statement of the *minimum* suction conditions required to prevent cavitation in a pump. The *required* or *minimum* NPSH must be determined by test and usually will be stated by the manufacturer. The *available* NPSH at installation must be at least equal to the required NPSH if cavitation is to be prevented. Increasing the available NPSH provides a margin of safety against the onset of cavitation. Figure 10.18 and the following symbols will be used to compute the NPSH:

p_a = absolute pressure in atmosphere surrounding gauge, Fig. 10.18.

p_s = gauge pressure indicated by gauge or manometer connected to pump suction at section *s-s*; may be positive or negative

p_t = absolute pressure on free surface of liquid in closed tank connected to pump suction

p_{vp} = vapor pressure of liquid being pumped corresponding to the temperature at section *s-s*. If the liquid is a mixture of hydrocarbons, p_{vp} must be measured by the *bubble point* method

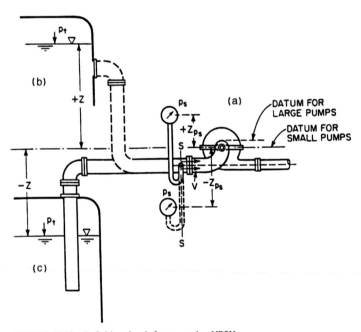

FIGURE 10.18   Definition sketch for computing NPSH.

h_f = lost head due to friction in suction line between tank and section *s-s*
V = average velocity at section *s-s*
Z, Z_{ps} = vertical distances defined by Fig. 10.18; may be positive or negative
γ = specific weight of liquid at pumping temperature

It is satisfactory to choose the datum for small pumps, as shown in Fig. 10.18. But with large pumps the datum should be raised to the elevation where cavitation is most likely to start. For example, the datum for a large horizontal-shaft propeller pump should be taken at the highest elevation of the impeller vane tips. The available NPSH is given by

$$h_{sv} = \frac{p_s - p_{vp}}{\gamma} + \frac{p_s}{\gamma} + Z_{ps} + \frac{V^2}{2g} \qquad (10.129)$$

or

$$h_{sv} = \frac{p_t - p_{vp}}{\gamma} + Z - h_f \qquad (10.130)$$

Consistent units must be chosen so that each term in Eqs. (10.129) and (10.130) represents feet (or meters) of the fluid pumped. Also,

$$h_{sv} = h_b + h_s \qquad (10.131)$$

Usually a positive value of h_s is called a *suction head* while a negative value of h_s is called a *suction lift*.

SCREW PUMPS AND MAXIMUM SUCTION LIFT

Suction lift occurs when the total available pressure at the pump inlet is below atmospheric pressure. It is normally the result of a change in elevation and pipe friction. Although screw pumps are capable of producing a high vacuum, it is not this vacuum that forces the fluid to flow. As previously explained, it is atmospheric or some other externally applied pressure that forces the fluid into the pump. Since atmospheric pressure at sea level corresponds to 14.7 psi absolute, or 30 inHg, this is the maximum amount of pressure available for moving the fluid, and suction lift cannot exceed these figures. In practice, a lower value of pressure is available; some of it is used up in overcoming friction in the inlet lines, valves, elbows, etc. It is considered the best practice to keep suction lift just as low as possible (Fig. 10.19).

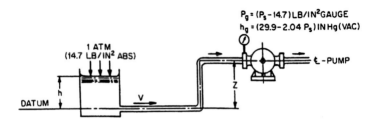

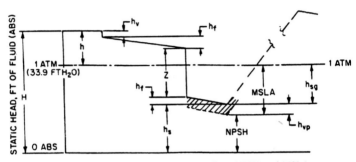

FIGURE 10.19 Relationship between hydraulic gradient, NPSH, and MSLA.

Total head at source = velocity head + elevation head
+ static head + friction head loss

$$H, \text{ft (m)} = h + \frac{33.9}{w} = h_v + Z + h_s + \Sigma(h_f)$$

$$= \frac{V^2}{2g} + Z + \frac{144P_s}{w} + \frac{144P_f}{w} \qquad (10.132)$$

Static head at pump inlet = net positive suction head
+ liquid vapor pressure, ft(m) abs

$$h_s = \text{NPSH} + h_{vp} \quad \text{or} \quad \text{NPSH} = h_s - h_{vp} \qquad (10.133)$$

Maximum suction lift available
= NPSH expressed in reference to atm (gauge reading)

$$\text{MSLA} = 1 \text{ atm} - \text{NPSH} \qquad (10.134)$$

where P_g, h_g = pressure gauge readings at pump inlet flange, psi gauge and
in Hg (vac)

P_s = absolute static pressure at pump inlet, psi

h_s, h_{sg} = static head at pump inlet, ft m^2 of liquid abs or gauge

Z = elevation head, ft (m) in reference to datum

h = reservoir liquid level, ft (m) in reference to datum

h_v, h_f = velocity head and friction head loss

P_{vp}, h_{vp} = liquid vapor pressure, psi abs, or heat, ft (m) abs

P_{sv} = net poritive inlet pressure psi abs

NPSH = net positive suction head, ft (m) of liquid abs

P_{svr} = net positive inlet pressure required by pump

MSLA = maximum suction lift available from pump, ft (m) of liquid or in Hg (vac)

w = specific weight of liquid

SPECIFIC SPEED OF COMMERCIAL PUMPS

Formulas below, Fig. 10.20, give typical specific speeds for commercial pumps. For any centrifugal pump, specific speed S is

$$S = \frac{\text{rpm}, \sqrt{\text{gpm}}}{h_{sv}^{3/4}} \tag{10.135}$$

For same Q, S, h_{sv}: $n_{\text{double suction}} = \sqrt{2}n_{\text{single suction}} = 1.414n_{\text{single suction}}$

For same Q, H, h_{sv}: $(n_s)_{\text{double suction}} = \sqrt{2}(n_s)_{\text{single suction}}$
$= 1.414\,(n_s)_{\text{single suction}}$

For same Q, n_s, h_{sv}: $H_{\text{double suction}} = 1.587 H_{\text{single suction}}$

For same Q, n: $(h_{sv})_{\text{double suction}} = 0.630(h_{sv})_{\text{single suction}}$

$$n_{\text{max}} = \frac{S(h_{sv})^{3/4}}{\sqrt{Q}} \cong \frac{8000(h_{sv})^{3/4}}{\sqrt{Q}}$$

Q = gpm; H = ft of fluid pumped; n = rpm; h_{sv} = ft of fluid pumped

Suction Specific Speed S

The *suction specific speed S* may be obtained by replacing H in Eq. 10.5 by h_{sv}

$$S = \frac{N\sqrt{Q}}{(h_{sv})^{3/4}} \tag{10.136}$$

Note that Q = one-half the discharge of a double-suction impeller when computing S. Equations in Fig. 10.20 may be combined to yield

$$\sigma = \left(\frac{n_s}{S}\right)^{4/3} \tag{10.137}$$

or
$$n_s = S(\sigma)^{1/4} \tag{10.138}$$

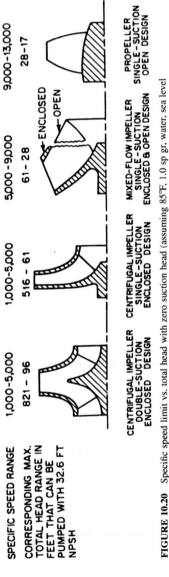

SPECIFIC SPEED RANGE	1,000-5,000	1,000-5,000	5,000-9,000	9,000-13,000
CORRESPONDING MAX. TOTAL HEAD RANGE IN FEET THAT CAN BE PUMPED WITH 32.6 FT NPSH	821 – 96	516 – 61	61 – 28	28–17
	CENTRIFUGAL IMPELLER DOUBLE-SUCTION ENCLOSED DESIGN	CENTRIFUGAL IMPELLER SINGLE-SUCTION ENCLOSED DESIGN	MIXED-FLOW IMPELLER SINGLE-SUCTION ENCLOSED & OPEN DESIGN	PROPELLER SINGLE-SUCTION OPEN DESIGN

FIGURE 10.20 Specific speed limit vs. total head with zero suction head (assuming 85°F, 1.0 sp gr, water, sea level equivalent to 32.6 ft NPSH 8,000 suction specific speed).

OPERATION OF CENTRIFUGAL PUMPS AT REDUCED FLOW RATES

When the flow through a centrifugal pump is reduced, the temperature rise in the liquid pumped can be quite rapid, and it can be computed from

$$T_m = \frac{42.4 \times P_{so}}{W_w \times C_w} \tag{10.139}$$

where T_m = temperature rise, °F/min
P_{so} = brake horsepower at shutoff
42.4 = conversion from bhp to Btu/min
W_w = net weight of liquid in pump, lb
C_w = specific heat of liquid (1.0 if liquid is water)

If flow is taking place through the pump, conditions become stabilized and it is possible to calculate the temperature rise through the pump for any given flow. Assuming that the liquid is water, the following formula can be used:

$$T = \frac{(\text{bhp} - \text{whp}) \times 2545}{\text{capacity, lb/h}} \tag{10.140}$$

where T = temperature rise, °F/min, and 2545 = Btu equivalent of 1 hp · hr. A more convenient formula relates the temperature rise to the total head and to the pump efficiency:

$$T = \frac{H}{778} \left(\frac{1}{e} - 1 \right) \tag{10.141}$$

where H = total head, ft, and e = pump efficiency at a given capacity. (Note that these formulas neglect the effect of the compressibility of the water. For a more exact calculation of temperature rise, especially when dealing with very high pressures, more precise thermodynamic calculations are indicated.)

EDUCTORS

Theory and Design

Eductor (Fig. 10.21) theory is developed from the Bernoulli equation. Static pressure at the entrance to the nozzle is converted to kinetic energy by permitting the fluid to flow freely through a converging-type nozzle. The resulting high-velocity stream entrains the suction fluid in the suction chamber, resulting in a flow of mixed fluids at an intermediate velocity. The

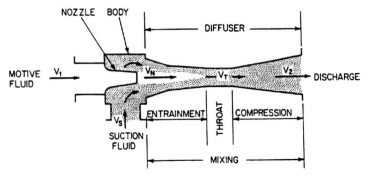

FIGURE 10.21 Jet nozzles convert pressure energy to velocity while diffusers entrain and mix the fluids and change velocity back to pressure.

diffuser section then converts the velocity pressure back to static pressure at the discharge of the eductor. Writing the Bernoulli equation for the motive fluid across the nozzle of an eductor, we have

$$\frac{P_1}{w_1} + \frac{V_1^2}{2g} = \frac{P_s}{w_1} + \frac{V_N^2}{2g} \qquad (10.142)$$

where P_1 = static pressure upstream, lb/ft^2
P_s = static pressure at suction (nozzle tip), lb/ft^2
V_1 = velocity upstream of nozzle, ft/s
V_N = velocity at nozzle orifice, ft/s
w_1 = specific weight of motive fluid, lb/ft^3

Upstream of the nozzle, all the energy is considered static head, so the velocity term V_1 drops out, yielding

$$\frac{V_N^2}{2g} = \frac{P_1 - P_s}{w_1} \qquad (10.143)$$

This term in called the *operating head*.

Across the diffuser, the same principle applies for the mixed fluid stream, except the effect is the reverse of a nozzle; hence,

$$\frac{P_s}{w_2} + \frac{V_T^2}{2g} = \frac{P_2}{w_2} + \frac{V_2^2}{2g} \qquad (10.144)$$

where P_s = static pressure at suction, lb/ft^2
P_2 = static pressure at discharge, lb/ft^2
V_T = velocity at diffuser throat, ft/s
V_2 = velocity downstream, ft/s
w_2 = specific weight of mixed fluids, lb/ft^2

At the discharge, it is assumed that all velocity head has been converted to static head; hence $V_2 = 0$ and

$$\frac{V_T^2}{2g} = \frac{P_2 - P_s}{w_2} \qquad (10.145)$$

This term is called the *discharge head*. The head ratio R_H is then defined as the ratio of the operating head to the discharge head:

$$R_H = \frac{V_N^2/2g}{V_T^2/2g} = \frac{V_N^2}{V_T^2} = \frac{(P_1 - P_s)/w_1}{(P_2 - P_s)/w_2} = \frac{(P_1 - P_s)w_2}{(P_2 - P_s)w_1} \qquad (10.146)$$

Since ratios are involved, it is convenient to replace specific weight with specific gravity.

$$R_H = \frac{(P_1 - P_s)\ \text{sp gr}_2}{(P_2 - P_s)\ \text{sp gr}_1} \qquad (10.147)$$

When both suction and motive fluid are the same, no gravity correction is required and Eq. (10.147) becomes

$$R_H = \frac{H_1 - H_s}{H_2 - H_s} \qquad (10.148)$$

where $H_1 - H_s$ = operating head, ft, and $H_2 - H_s$ = discharge head, ft. Entrainment conditions are defined by the basic momentum equation

$$M_1 V_N + M_s F_s = (M_1 + M_s)V_T \qquad (10.149)$$

where M_1 = mass of motive fluid, slugs
M_s = mass of suction fluid, slugs
V_N = velocity at nozzle discharge, ft/s
V_s = velocity at suction inlet, ft/s
V_T = velocity at diffuser throat, ft/s

The velocity of approach at the suction inlet is zero; therefore

$$M_s = M_1 \left(\frac{V_N}{V_T} - 1\right) \qquad (10.150)$$

and the term below is defined as the *weight operating ratio:*

$$R_w = \frac{M_s}{M_1} = \frac{V_N}{V_T} - 1 \qquad (10.151)$$

Observe that the term V_N^2/V_T^2 has previously been defined as the head ratio R_H; therefore

$$R_w = \sqrt{R_H} - 1 \qquad (10.152)$$

The volume ratio R_q is then simply

$$\frac{Q_s}{Q_1} = R_w \frac{\text{sp gr}_1}{\text{sp gr}_2} \qquad (10.153)$$

where Q_s = suction flow in volumetric units and Q_1 = motive flow in volumetric units.

PUMP SHAFT DESIGN

Design Criteria

Torsional Stress. The torsional stress in the shafting may be calculated by the following equations:

$$S_s = \frac{16T}{D_o^2} \qquad \text{for solid shafting} \qquad (10.154)$$

$$S_s = \frac{16T}{D_o^2 (1 - D^4/D_o^4)} \qquad \text{for tubular shafting} \qquad (10.155)$$

where S_s = torsional shear stress, lb/in^2
   T = transmitted torque, lb·lb
   D = shaft inner diameter (tubular shaft only), in
   D_o = shaft outer diameter, in

The allowable value of shear stress depends upon the material being used and whether it is subjected to other loads, such as bending or compression. The design safety factor on the shafting should be equal to or greater than the other components in the drivetrain.

Critical Speed. The critical speed of a driveshaft is determined by the deflection, or "sag," of the shaft in a horizontal position under its own weight. The less the sag, the higher the critical speed. In practical terms, a long, slender shaft will have a low critical speed while a short, large-diameter shaft will have a very high critical speed. Calculation of the deflection of a simply supported shaft is as follows:

$$y = \frac{5wL^4}{384EI} = \text{shaft deflection} \qquad (10.156)$$

noting that

$$I = \frac{\pi D_o^4}{64} \qquad \text{for solid shafting} \qquad (10.157)$$

$$I = \frac{\pi(D_o^4 - D^4)}{64} \qquad \text{for turbular shafting} \qquad (10.158)$$

where w = weight of shaft per unit length, lb/in
     L = length between bearing supports, in
     E = Young's modulus, psi
     I = moment of inertia, in⁴

Given the natural deflection of the shaft, it is possible to calculate the first critical speed from the equation

$$N_{\text{crit}} = 187 \sqrt{\frac{1}{y}} \qquad (10.159)$$

which expresses critical speed directly in rpm of the rotating shaft.

PUMP POWER CALCULATIONS

Power Output

The water horsepower (whp) or useful work done by the pump is found by

$$\text{whp} = \frac{\text{lb of liquid pumped/min} \times \text{total head in ft of liquid}}{33,000} \qquad (10.160)$$

If the liquid has a specific gravity of 1 and the specific weight of the liquid is 62.3 lb/ft² at a temperature of 68°F, the formula is

$$\text{whp} = \frac{\text{gpm} \times \text{head, ft}}{3960} \qquad (10.161)$$

Power Input

The brake horsepower required to drive the pump is found by the formula

$$\text{bhp} = \frac{\text{gpm} \times \text{total head, ft}}{3960 \times \text{pump efficiency}} \qquad (10.162)$$

where pump efficiency is obtained by the formula

$$\text{Pump efficiency} = \frac{\text{output}}{\text{input}} = \frac{\text{whp}}{\text{bhp}} \qquad (10.163)$$

The electric horsepower (ehp) input to the motor is

$$\text{ehp} = \frac{\text{bhp}}{\text{motor efficiency}} \qquad (10.164)$$

$$= \frac{\text{gpm} \times \text{head, ft}}{3960 \times \text{pump efficiency} \times \text{motor efficiency}} \qquad (10.165)$$

The kilowatt input to the motor is

$$\text{kW input} = \frac{\text{bhp} \times 0.746}{\text{motor efficiency}} \qquad (10.166)$$

$$= \frac{\text{gpm} \times \text{head} \times 0.746}{3960 \times \text{pump efficiency} \times \text{motor efficiency}} \qquad (10.167)$$

Pump Efficiency

The pump efficiency is found by

$$\text{Pump efficiency} = \frac{\text{output}}{\text{input}} = \frac{\text{whp}}{\text{bhp}} \qquad (10.168)$$

For an electric motor-driven pumping unit, the overall efficiency is found by

$$\text{Overall efficiency} = \text{pump efficiency} \times \text{motor efficiency} \qquad (10.169)$$

In many specifications it is required that the actual job motor be used to drive its respective pump during shop or field testing. By using this test setup, the overall efficiency then becomes what is commonly called *wire-to-water* efficiency, which is expressed by

$$\text{Overall efficiency} = \frac{\text{whp}}{\text{ehp input}} = \frac{\text{whp}}{\text{ehp}} \qquad (10.170)$$

HOW TO FIGURE STEEL-PIPE SCHEDULE

Use Eq. (10.171) to figure the pipe thickness needed, material, and nominal pipe size for a particular job.

$$t_m = \frac{PD}{2S + 2yP} + C \qquad (10.171)$$

where t_m = minimum wall thickness, in. If solving for nominal thickness, add 12½ percent to cover mill tolerance

P = maximum internal pressure, psig, at operating temperature

D = outside pipe diameter, in

S = allowable stress at operating temperature, psi, from engineering handbooks

C = allowance for structural stability only; make additional allowance for corrosion in table below

Threaded pipe (¾-in and smaller)	0.065
Threaded pipe (1-in and larger)	depth of thread
Grooved pipe	depth of groove
Plain-end pipe (3½-in and smaller)	0.065
Plain-end pipe (4-in and larger)	0.000

y = metal and temperature coefficient:

	Temperature, °F					
Steel type	Under 900	950	1000	1050	1100	Above 1150
Ferritic	0.4	0.5	0.7	0.7	0.7	0.7
Austenitic	0.4	0.4	0.4	0.4	0.5	0.7

Now check a table of pipe properties and pick a schedule of pipe having next-larger wall thickness. Recheck the pressure and temperature rating, using actual pipe dimensions. Using actual pipe ID, recheck hydraulic losses.

FIGURE PIPE WEIGHT PER FOOT OF LENGTH

First get the inside diameter in inches:

$$ID = OD - 2t \qquad (10.172)$$

where ID = inside diameter of pipe, in

OD = outside diameter from table, in

t = pipe wall thickness from table, in

Then

$$\text{Wt/ft of length} = 0.85 \ \pi(OD^2 - ID^2) \qquad (10.173)$$

FORMULAS FOR NONLAMINAR (TURBULENT) AND LAMINAR FLOW IN PIPES

Here are 21 formulas—6 for nonlaminar (turbulent) flow and 15 for laminar flow of fluids in pipes (Table 10.3). These formulas allow easy conversion when different variables are known for the flow situation. The variables involved are pipe length, fluid velocity, pipe diameter, pressure loss, flow rate, friction factor, and absolute viscosity.

Nomenclature

$cSt = cP \ (lb \cdot s/ft^2)$

$d =$ pipe inside diameter in

$D =$ pipe inside diameter ft

$f =$ friction factor, dimensionless

$g =$ gravity constant, 32.2 ft/s^2

$H_L =$ head loss, ft

$L =$ pipe length, ft

$L_{in} =$ pipe length, in

$\Delta p =$ pressure loss, psi

$\Delta P =$ pressure loss, psf

$q =$ flow, gal/min

$Q =$ flow, ft^3/s

$Q_{in^3} =$ flow, in^3/s

$R_e =$ Reynolds' number $\rho DV/\mu$. Typical units are:
 3162 q/dv_{cSt} gal/min, in, cSt
 50.6 $q\gamma/d\mu_{cp}$ gal/min, lb/ft^3, in, cP
 $\gamma DV/g\mu$ lb/ft^3, ft, ft/s, lb·s/ft^2

$v =$ fluid velocity, in/s

$V =$ fluid velocity, ft/s

$\mu =$ absolute viscosity, lb·s/ft^2

$\nu = \mu/\rho$, kinematic viscosity, ft^2/s

$\rho =$ mass density, lb·s^2/ft^4 = slugs/ft^3

$\gamma =$ weight density, lb/ft^3

$\epsilon/D =$ relative roughness of pipe wall. Values of ϵ, ft: drawn tubing, 5 $\times$ 10^{-6}; steel or wrought iron, 150 $\times$ 10^{-6}. $D =$ inside diameter ft.

TABLE 10.3 Formulas for Flow in Pipes

General empirical relationships: all flows, $H_L = f \dfrac{L}{D} \dfrac{V^2}{2g}$; laminar only, $H_L = 32 \dfrac{\mu L V}{\rho g d^2}$; useful conversions, for oil only (= 55 lb/ft²).

Variables							Formulas		
L	V	D	ΔP	Q	f	μ	Pressure loss	Velocity	Flow
							All flows		
ft	$\dfrac{\text{ft}}{\text{s}}$	in	psi	$\dfrac{\text{ft}^3}{\text{s}}$	—	—	$\Delta p = f \dfrac{L}{d} V^2 \times 0.072$	$V = \sqrt{\dfrac{d\,\Delta p}{fL}} \times 3.73$	$Q = \sqrt{\dfrac{d^5\,\Delta p}{fL}} \times 0.0203$
ft	$\dfrac{\text{ft}}{\text{s}}$	in	psi	gal/min (q)	—	—	$\Delta p = f \dfrac{L}{d^5} q^2 \times 0.0123$	$V = \sqrt{\dfrac{d\,\Delta p}{fL}} \times 3.73$	$q = \sqrt{\dfrac{d^5\,\Delta p}{fL}} \times 9.1$
							Laminar only		
ft	$\dfrac{\text{ft}}{\text{s}}$	in	psi	$\dfrac{\text{ft}^3}{\text{s}}$	—	cSt	$\Delta p = \nu \dfrac{L}{d^2} V \times 0.0006$	$V = \dfrac{d^2\,\Delta p}{\nu L} \times 1670$	$Q = \dfrac{d^4\,\Delta p}{\nu L} \times 9.1$
ft	$\dfrac{\text{ft}}{\text{s}}$	in	psi	gal/min	—	cSt	$\Delta p = \nu \dfrac{L}{d^4} q \times 2.45 \times 10^{-4}$	$V = \dfrac{d^2\,\Delta p}{\nu L} \times 1670$	$q = \dfrac{d^4\,\Delta p}{\nu L} \times 4080$
ft	$\dfrac{\text{in}}{\text{s}}$	in	psi	$\dfrac{\text{in}^3}{\text{s}}$	—	$\dfrac{\text{lb} \cdot \text{s}}{\text{in}^2}$	$\Delta p = \mu \dfrac{L_{\text{in}}}{d^4} Q_{\text{in}^3} \times 40.75$	$V = \dfrac{d^2\,\Delta p}{\mu L_{\text{in}}} \times 0.0312$	$Q_{\text{in}^3} = \dfrac{d^4\,\Delta p}{\mu L_{\text{in}}} \times 0.0245$
ft	$\dfrac{\text{ft}}{\text{s}}$	ft	lb/ft²	$\dfrac{\text{ft}^3}{\text{s}}$	—	$\dfrac{\text{lb} \cdot \text{s}}{\text{ft}^2}$	$\Delta p = \mu \dfrac{L}{D^2} V \times 32$	$V = \dfrac{D\,\Delta P}{\mu L} \times 0.0312$	$Q = \dfrac{D^4\,\Delta P}{\mu L} \times 0.0248$
ft	$\dfrac{\text{ft}}{\text{s}}$	ft	lb/ft²	gal/min	—	$\dfrac{\text{lb} \cdot \text{s}}{\text{ft}^2}$	$\Delta p = \mu \dfrac{L}{D^4} q \times 0.091$	$V = \dfrac{D^2\,\Delta P}{\nu L} \times 0.0312$	$q = \dfrac{D^4\,\Delta P}{\nu L} \times 11$

FORMULAS FOR AIRFLOW IN PIPES, VALVES, AND FITTINGS

Manufacturers publish a variety of airflow formulas. Here are 10 such and their defining data, along with conversion formulas (Tables 10.4 to 10.6). Any of these formulas, used consistently in a design, will give satisfactory results.

Nomenclature

Q = airflow in standard units, scfm (14.7 psi, 68°F)

q = airflow at actual conditions, cfm. $Q = q(P/14.7)(528/T)$

V = velocity, ft/s (average through valve)

P = pressure in absolute units, psia (subscript D = downstream, U = upstream)

p = gauge pressure, psi

ΔP = pressure drop, psi

r = pressure ratio P_D/P_U

ρ = density, lb/ft^3

G = specific gravity, ρ_{gas}/ρ_{air}

T = absolute temperature, °R = °F + 460

A = inlet pipe area, in^2

D_e = diameter of equivalent sharp-edge orifice, in (coefficient of discharge $C_D = 0.6$)

M = molecular weight, lb ($M = 29$ lb for air)

TABLE 10.4 Manufacturer Information for Airflow Instruments

Component	Flow coefficient	Defining equation
Hand valve	$C_V = 1.26$	$C_V = \dfrac{Q \times 60}{1360} \sqrt{\dfrac{GT_U}{\Delta P \times P_U}}$
Pressure reducer	$D_o = 0.25$	$D_o = \sqrt{\dfrac{Q}{33P_U} \times \dfrac{1}{\sqrt{r(r^{0.43} - r^{0.71})}}}$
Control valve	F = unknown	$F = \dfrac{Q}{P_U \sqrt{8/5}} \sqrt{\dfrac{1}{r(1 - r)(3 - r)}}$
Air motor	$Q = 250$ scfm; $P = 600$ psia (given)	

TABLE 10.5 Conversion Formulas for Airflow Variables

	D_o	C_v	F	K		
				$r = 1.0$	$r = 0.75$	$r = 0.5$
D_o		$= 0.236\sqrt{C_v}$	$= 0.316\sqrt{F}$	$= 1.456\dfrac{\sqrt{A}}{K^{1/4}}$	$= 1.521\dfrac{\sqrt{A}}{K^{1/4}}$	$= 1.641\dfrac{\sqrt{A}}{K^{1/4}}$
C_v	$= 18.0 D_o^2$		$= 1.8F$	$= 38.2\dfrac{A}{\sqrt{K}}$	$= 41.5\dfrac{A}{\sqrt{K}}$	$= 48.3\dfrac{A}{\sqrt{K}}$
F	$= 10 D_o^2$	$= 0.556 C_v$		$= 21.2\dfrac{A}{\sqrt{K}}$	$= 23.1\dfrac{A}{\sqrt{K}}$	$= 26.9\dfrac{A}{\sqrt{K}}$
$r = 1.0$	$= 4.5\dfrac{A^2}{D_o^4}$	$= 1460\dfrac{A^2}{C_v^2}$	$= 450\dfrac{A^2}{F^2}$			
$K, r = 0.75$	$= 5.36\dfrac{A^2}{D_o^4}$	$= 1725\dfrac{A^2}{C_v^2}$	$= 534\dfrac{A^2}{F^2}$			
$r = 0.5$	$= 7.29\dfrac{A^2}{D_o^2}$	$= 2330\dfrac{A^2}{C_v^2}$	$= 724\dfrac{A^2}{F^2}$			

Note: The K factor varies with r and A and you must know which values the manufacturer used to derive its published K. For example, if K was derived at $r = 0.75$ and valve inlet pipe area $A = 0.2$, then $F = 23.1 \times 0.2/\sqrt{K} = 4.62/\sqrt{K}$.

TABLE 10.6 Typical Airflow Formulas

Formula for flow (subcritical) (Q = scfm, standard ft³/min)	Flow coefficient defined	
$Q = 33 D_o^2 P_t \sqrt{r(r^{0.43} - r^{0.71})}$	D_o = Equivalent sharp-edged orifice (coeff. of discharge $C_D = 0.6$)	$= \sqrt{\dfrac{Q}{33\,P_U}} \times \sqrt{\dfrac{1}{\sqrt{r^{0.43} - r^{0.71}}}}$
$Q = \dfrac{963}{60} C_V \sqrt{\dfrac{\Delta P(P_U + P_D)}{GT_U}}$	C_V = valve coefficient	$= \dfrac{W \times 60}{963} \sqrt{\dfrac{GT_U}{\Delta P(P_U + P_D)}}$
$Q = \dfrac{1360}{60} C_V \sqrt{\dfrac{\Delta P \times P_U}{GT_U}}$	C_V = flow coefficient	$= \dfrac{Q \times 60}{1360} \sqrt{\dfrac{GT_U}{\Delta P \times P_U}}$
$Q = \dfrac{1390}{60} C_V \sqrt{\dfrac{\Delta P \times P_D}{GT_U}}$	C_V = capacity factor	$= \dfrac{Q \times 60}{1390} \sqrt{\dfrac{GT_U}{\Delta P \times P_D}}$
$Q = \dfrac{5180}{60} C_V \sqrt{\dfrac{P_U^2 - P_D^2}{MT_U}}$	C_V = valve flow coefficient	$= \dfrac{Q \times 60}{5180} \sqrt{\dfrac{MT_U}{P_U^2 - P_D^2}}$
$Q = \dfrac{963}{60} C_V \sqrt{\dfrac{P_V^2 - P_D^2}{GT_U}}$	C_V = flow coefficient	$= \dfrac{Q \times 60}{963} \sqrt{\dfrac{GT_U}{P_U^2 - P_D^2}}$

$$Q = \frac{2.32^{0.443}}{60} \, C_a \, \frac{\Delta P^{0.443} \times P_U^{0.6}}{\sqrt{GT_U/520}} \qquad C_o = \text{gas flow coefficient} \qquad = \frac{Q \times 60}{(2.32)^{0.443}} \times \frac{\sqrt{GT_U/520}}{\Delta P^{0.443} \times P_U^{0.6}}$$

$$Q = FP_U \sqrt{\frac{4}{3}} \sqrt{1 - r^2} \qquad F = \text{NBS flow factor (\#1)} \qquad = \frac{Q}{P_U \sqrt{4/3}} \sqrt{\frac{1}{1 - r^2}}$$

$$Q = FP_U \sqrt{\frac{8}{5}} \sqrt{r(1 - r)(3 - r)} \qquad F = \text{NBS flow factor (\#2—better)} \qquad = \frac{Q}{P_U \sqrt{8/5}} \sqrt{\frac{1}{r(1 - r)(3 - r)}}$$

$$Q = 38.1 \, P_U A \sqrt{\frac{1 - r}{K}} \qquad K = K \text{ factor} \qquad = \frac{2g}{\rho V^2} \, \Delta P$$

Taken from catalogs.

281

W = water weight flow, lb/s

C_v, K, F, D_e = typical flow coefficients (also called flow constants and flow factors) in a flow equation

AIR INTAKE SYSTEMS FOR INTERNAL COMBUSTION ENGINES

The first step in selecting and designing an air intake system for internal combustion engines is to determine the quantity of air required for combustion. The quantity of air required for combustion, in cubic feet per minute, is

$$\text{Required air} = \frac{B^2 S \times \text{rpm} \times N}{2200K} \qquad (10.174)$$

where B = cylinder bore, in
S = piston stroke, in
rpm = engine speed, rpm
N = number of cylinders
K = constant: two-cycle, 1; four-cycle, 4

Volumetric efficiency will vary with engine design, but an average of 80 percent may be used to determine the air cleaner size.

FOUNDATIONS FOR INTERNAL COMBUSTION ENGINES

Foundation design is extremely important for internal combustion engine intallations. The size and mass of the foundation depend on the dimensions and weight of the engine and generator, or pump if it is a pump drive installation. The following minimum standards should be followed:

1. Width should exceed the equipment width and length by a minimum of 1 ft.
2. The depth should be sufficient to provide a weight equal to 1.3 to 1.5 times the weight of the equipment. This depth may be determined by

$$H = \frac{(1.3 - 1.5)W}{LB \times 135} \qquad (10.175)$$

where H = depth of foundation, ft
L = length of foundation, ft
B = width of foundation, ft

135 = density of concrete, lb/ft^3
W = weight of equipment

3. The soil-bearing load should not exceed the building standard codes. It may be calculated by

$$\text{Bearing load} = \frac{(2.3 - 2.5)W}{BL} \qquad (10.176)$$

where W = weight of equipment
B = width of foundation, ft
L = length of foundation, ft

INDEX

ABOUT THE AUTHOR

Tyler Hicks is a registered professional engineer associated with International Engineering Associates. He is an active member of ASME, the U.S. Naval Institute, and the American Merchant Marine Museum Foundation. Mr. Hicks is the author of McGraw-Hill's *Handbook of Mechanical Engineering Calculations* and *Standard Handbook of Engineering Calculations*.